I0464504

THE SCIENCE OF DYSLEXIA

HEARING

BEFORE THE

COMMITTEE ON SCIENCE, SPACE, AND TECHNOLOGY
HOUSE OF REPRESENTATIVES

ONE HUNDRED THIRTEENTH CONGRESS

SECOND SESSION

September 18, 2014

Serial No. 113–95

Printed for the use of the Committee on Science, Space, and Technology

Available via the World Wide Web: http://science.house.gov

U.S. GOVERNMENT PUBLISHING OFFICE

92–328PDF   WASHINGTON : 2015

For sale by the Superintendent of Documents, U.S. Government Publishing Office
Internet: bookstore.gpo.gov   Phone: toll free (866) 512–1800; DC area (202) 512–1800
Fax: (202) 512–2104   Mail: Stop IDCC, Washington, DC 20402–0001

COMMITTEE ON SCIENCE, SPACE, AND TECHNOLOGY

HON. LAMAR S. SMITH, Texas, *Chair*

DANA ROHRABACHER, California
RALPH M. HALL, Texas
F. JAMES SENSENBRENNER, JR.,
  Wisconsin
FRANK D. LUCAS, Oklahoma
RANDY NEUGEBAUER, Texas
MICHAEL T. McCAUL, Texas
PAUL C. BROUN, Georgia
STEVEN M. PALAZZO, Mississippi
MO BROOKS, Alabama
RANDY HULTGREN, Illinois
LARRY BUCSHON, Indiana
STEVE STOCKMAN, Texas
BILL POSEY, Florida
CYNTHIA LUMMIS, Wyoming
DAVID SCHWEIKERT, Arizona
THOMAS MASSIE, Kentucky
KEVIN CRAMER, North Dakota
JIM BRIDENSTINE, Oklahoma
RANDY WEBER, Texas
CHRIS COLLINS, New York
BILL JOHNSON, Ohio

EDDIE BERNICE JOHNSON, Texas
ZOE LOFGREN, California
DANIEL LIPINSKI, Illinois
DONNA F. EDWARDS, Maryland
FREDERICA S. WILSON, Florida
SUZANNE BONAMICI, Oregon
ERIC SWALWELL, California
DAN MAFFEI, New York
ALAN GRAYSON, Florida
JOSEPH KENNEDY III, Massachusetts
SCOTT PETERS, California
DEREK KILMER, Washington
AMI BERA, California
ELIZABETH ESTY, Connecticut
MARC VEASEY, Texas
JULIA BROWNLEY, California
ROBIN KELLY, Illinois
KATHERINE CLARK, Massachusetts

CONTENTS

September 18, 2014

Page

Appendix I: Answers to Post-Hearing Questions

Appendix II: Additional Material for the Record

THE SCIENCE OF DYSLEXIA

THURSDAY, SEPTEMBER 18, 2014

House of Representatives,
Committee on Science, Space, and Technology,
Washington, D.C.

The Committee met, pursuant to call, at 11:21 a.m., in Room 2318 of the Rayburn House Office Building, Hon. Lamar Smith [Chairman of the Committee] presiding.

LAMAR S. SMITH, Texas
CHAIRMAN

EDDIE BERNICE JOHNSON, Texas
RANKING MEMBER

Congress of the United States
House of Representatives

COMMITTEE ON SCIENCE, SPACE, AND TECHNOLOGY

2321 RAYBURN HOUSE OFFICE BUILDING

WASHINGTON, DC 20515-6301

(202) 225-6371
www.science.house.gov

The Science of Dyslexia

Thursday, September 18, 2014
11:00 a.m. to 1:00 p.m.
2318 Rayburn House Office Building

Witnesses

Panel 1:

Hon. Bill Cassidy, Member, U.S. House of Representatives

Hon. Julia Brownley, Member, U.S. House of Representatives

Panel 2:

Dr. Sally Shaywitz, Audrey G. Ratner Professor in Learning Development, Yale University
School of Medicine and Co-Director, Yale Center for Dyslexia and Creativity, Yale University

Mr. Max Brooks, Author and Screenwriter

Ms. Stacy Antie, Parent and Advocate

Dr. Peter Eden, President, Landmark College

Dr. Guinevere Eden, Director, Center for the Study of Learning (CSL) and Professor,
Department of Pediatrics, Georgetown University Medical Center

U.S. HOUSE OF REPRESENTATIVES
COMMITTEE ON SCIENCE, SPACE, AND TECHNOLOGY
FULL COMMITTEE

HEARING CHARTER

The Science of Dyslexia

Thursday, September 18, 2014
11:00 a.m. - 1:00 p.m.
2318 Rayburn House Office Building

Purpose

The Committee on Science, Space, and Technology will hold a hearing entitled *The Science of Dyslexia* on Thursday, September 18, 2014, in Room 2318 Rayburn House Office Building. Dyslexia is a difficulty to read fluently and with accurate comprehension despite a normal or above-average intelligence. It is the most common learning disability, with an estimated 1 in 5 persons suffering from some form of dyslexia.[1] While dyslexia is considered a learning disability, many talented people—especially in science, engineering, and the creative arts—have been diagnosed with dyslexia, including Albert Einstein, Thomas Edison, and John Chambers, CEO of Cisco Systems.[2][3]

The purpose of this hearing is to understand the latest scientific research in dyslexia, to discuss promising future research directions and promising treatments for people with dyslexia to overcome challenges they face, and to explore educational opportunities for students with dyslexia in fields of science, technology, engineering, and mathematics (STEM). Witnesses will testify on their personal experience with dyslexia or how they helped others overcome this challenge through innovative and creative problem-solving.

Witnesses

Panel 1:
- **Hon. Bill Cassidy**, Co-Chair of Bipartisan Congressional Dyslexia Caucus
- **Hon. Julia Brownley**, Co-Chair of Bipartisan Congressional Dyslexia Caucus

Panel 2:
- **Dr. Sally Shaywitz**, Professor, Yale Center for Dyslexia and Creativity, Yale University
- **Mr. Max Brooks**, Author and Screenwriter
- **Ms. Stacy Antie**, Parent and Advocate
- **Dr. Peter Eden**, President, Landmark College
- **Dr. Guinevere Eden**, Director, Center for the Study of Learning (CSL) and Professor, Department of Pediatrics, Georgetown University Medical Center

---

[1] http://www.yalescientific.org/2011/04/the-paradox-of-dyslexia-slow-reading-fast-thinking/
[2] http://www.dyslexia.com/famous.htm
[3] http://www.businessinsider.com/cisco-ceo-john-chambers-talks-dyslexia-2014-7

Hearing Overview

Introduction

Dyslexia, a developmental reading disorder, is characterized by difficulty with learning to read fluently and with accurate comprehension despite normal or above-average intelligence. This language processing disorder can hinder reading, writing, spelling and sometimes even speaking. Dyslexia is the most common learning difficulty and most recognized reading disorder. Notable scientists who had dyslexia through history include Leonardo Da Vinci, Albert Einstein, Nikola Tesla and James Clerk Maxwell.[4]

One out of every five people struggle with dyslexia.[5] Unfortunately many of these individuals remain undiagnosed, untreated and struggling with the impact of their dyslexia. Dyslexia can affect people differently and depends upon the severity of the learning disability and the success of alternate learning methods. For example, some individuals may have trouble only with reading and spelling, while others struggle to write. Some children may show few signs of difficulty with early reading and writing. However, as adults, they may have trouble with complex language skills, such as grammar, reading comprehension and more in-depth writing. Furthermore, adults with unidentified dyslexia often work in jobs below their intellectual capacity.

People with dyslexia are often very creative, and they often think of unexpected ways to solve a problem or tackle a challenge. Researchers do not understand whether this creativity comes from the extra work dyslexics must do to succeed at reading, or whether dyslexics are just naturally creative. However, many individuals with dyslexia have exceeded expectations and having successful careers.

Issues for Consideration

The exact causes of dyslexia are not completely understood, but brain imaging studies show differences in the structure[6] and function[7] of the brains of dyslexic people. Moreover, most people with dyslexia have been found to have problems with identifying the separate speech sounds within a word and/or learning how letters represent those sounds, a key factor in their reading difficulties. Dyslexia is not due to either a lack of intelligence or desire to learn; with appropriate teaching methods, dyslexics can learn successfully.

Assistive technology offers a way for dyslexics to save time and overcome some of the issues they may encounter because of their dyslexia, such as slow note-taking or unreadable handwriting, and allows them to use their time productively where they are gifted. For dyslexic

---

[4] http://en.wikipedia.org/wiki/List_of_people_diagnosed_with_dyslexia
[5] http://dyslexia.yale.edu/MDAI/
[6] Krafnick, A.J., Flowers, D.L., Luetje, M.M., Napoliello, E.M. and Eden, G.F., An investigation into the origin of anatomical differences in dyslexia, Journal of Neuroscience, 34(3): 901-8, 2014. doi:10.1523/JNEUROSCI.2092-13.2013. PMCID:PMC3891966
[7]Eden, G. F., Jones, K.M., Cappell, K., Gareau, L., Wood, F.B., Zeffiro, T.A., Dietz, N.A.E., Agnew, J.A. and Flowers, D.L., Neural changes following remediation in adult developmental dyslexia, Neuron, 44(3): 411-422, 2004. doi:10.1016/j.neuron.2004.10.019

students, technology opens doors and allows them to demonstrate their knowledge in ways that were unimaginable in the past. Reading is the area in which students with dyslexia struggle the most. Fortunately, many mobile apps are available that can be of help.[8]

It is difficult to find a job today which does not require some level of reading, writing and memory, or some use of a computer. Adults with dyslexia sometimes struggle with time management and organization at work. Planning and organizing, setting out timetables, distinguishing between the important and the urgent, remembering appointments, passing on telephone messages from memory and meeting deadlines can be exceptionally difficult for people with dyslexia. Individuals may get bogged down, overwhelmed by the workload and perform poorly. [9]

Initial job training would be maximized by taking into account the specific needs of employees with dyslexia. This requires flexibility in the approach to training, provision of information in alternative formats, multi-sensory learning techniques, more time and repetition of information when necessary.[10]

---

[8] http://www.ncld.org/students-disabilities/assistive-technology-education/apps-students-ld-dyslexia-reading-difficulties
Apps to Help Students with Dyslexia and Reading Difficulties: http://ncld.org/students-disabilities/assistive-technology-education/apps-students-ld-dyslexia-reading-difficulties
Assistive Tech Innovations: 14 New Apps & Other Tools: http://ncld.org/students-disabilities/assistive-technology-education/apps-assistive-technology
Apps for Students with LD: Organization and Study: http://ncld.org/students-disabilities/assistive-technology-education/apps-students-ld-organization-study
An Overview of Assistive Technology: http://ncld.org/students-disabilities/assistive-technology-education/overview-assistive-technology

[9] http://www.ncld.org/adults-learning-disabilities/jobs-employment-ld/strengths-challenges-workplace
[10] http://ncld.org/adults-learning-disabilities/jobs-employment-ld/common-problems-easy-solutions

Chairman SMITH. The Committee on Science, Space, and Technology will come to order. Good morning to everyone, and welcome to today's hearing titled "The Science of Dyslexia," and I want to say at the outset, it is nice to see such interest. This is one of the best attended hearings we have had, and there is almost a festive atmosphere in the room, and I think that is because we all hope some good will come out of this hearing. There is certainly a common bond that unites all of us in this room as well, and we have no adverse witnesses, so it all adds up to what we expect to be a very productive hearing.

And before we start, too, and before I recognize myself and the Ranking Member for an opening statement, as much work and effort has gone into this hearing as I think with any other hearing we have possibly had, and I want to thank those who have done so much and worked so hard to get us all here and to focus on such an important subject, so I want to recognize our Chief of Staff, Jennifer Brown, to my left, Chris Shank to my left over here, Richard Yamada, who is sitting next to me, Kirsten Duncan and Christian Rice as well, who I think are to the left here too, but they have put in a great amount of time and effort, and we thank them for their contributions.

Well, welcome everyone to today's hearing on the science of dyslexia. One out of every five people struggle with dyslexia in its various forms. In fact, it is the most common reading disability in America. Yet many Americans remain undiagnosed, untreated, and silently struggle with school or work.

People with dyslexia think in a way that others do not. But typically in our school systems today, there is not recognition, early detection, or enough teachers who are trained to spot symptoms of dyslexia early enough to get the students the intervention they need.

That is why we have recently seen grassroots groups, like Decoding Dyslexia, form nationwide and more specialized schools started to fill the gap. Unfortunately, not everyone has access to these types of schools and the learning strategies they instill in their students to help them become successful.

I hope today's hearing will serve two purposes: first, contribute to our understanding as policymakers about the neuroscience of dyslexia, and secondly, build awareness of dyslexia's effect on those of all ages if we fail to diagnose it.

Some may ask why the Science Committee chooses to tackle the issue of dyslexia. My response is simple: many scientists, innovators and other outside-the-box thinkers are dyslexic, such as Albert Einstein, Leonardo da Vinci, and Galileo, to name a few. Many who have dyslexia have used their unique outlook on the world to their advantage. Filmmakers, actors and entertainers such as Steven Spielberg, Henry Winkler, and Jay Leno use their gift to create one-of-a-kind entertainment for us all to enjoy.

In modern times, Dr. Carol Greider of the Johns Hopkins School of Medicine, who won the Nobel Prize in 2009, has dyslexia. John Chambers, the long-time CEO of Cisco Systems, also has dyslexia. In a recent interview, Chambers spoke about his struggles: "It would surprise you how many government and business leaders have dyslexia. Some people view it as a weakness and maybe it is, but because of my weakness I have learned other ways to accomplish the same goal with faster speed. So in math, I can do equations faster by eliminating the wrong answers quicker than I can get the right answer. It is one of the reasons I talk to young people with dyslexia pretty regularly. You have to have role models.''

We need to unleash the intelligence of people with dyslexia, like Einstein, da Vinci, Carol Greider, and John Chambers. We cannot afford for young, talented students not to reach their potential.

And I am glad to see the National Science Foundation fund studies in how astrophysicists with dyslexia view the universe differently due to the visual-spatial skills common in dyslexics. In fact, Matt Mountain, the Lead Astronomer and Director of the Hubble Space Telescope Science Institute, has dyslexia, and without an objection, we will insert his testimony into the record at this point.

[The information appears in Appendix II]

Chairman SMITH. Also, the National Institutes of Health is studying the neuroscience of dyslexia, including the work of our witnesses, Dr. Sally Shaywitz and Dr. Guinevere Eden, as well as funding studies on how dyslexic students can best learn. Beyond the research, we will hear from someone with dyslexia, the parent of a dyslexic student, and an educator for those with learning disabilities like dyslexia.

I have a personal connection with dyslexia since my niece is dyslexic. And a favorite, young 10-year-old named Leighton, a young friend who has dyslexia, has been with me on a Texas ranch. He may be challenged by language arts but he makes up for it with perfect eyesight and exceptional accuracy with his BB gun. And you don't want to compete with him playing Minecraft on his iPad.

Over 80 Members of Congress have joined the bipartisan Congressional Dyslexia Caucus co-chaired by Representative Bill Cassidy and Science Committee Member Julia Brownley. I thank them both for their work in helping educate the public about dyslexia and for advocating policies that support those individuals who have dyslexia.

I also want to acknowledge one of my constituents, Robbi Cooper, who is to my left, who traveled from Austin, Texas, to be here today. She has shared many stories with me about her son Ben who has dyslexia.

More parents as well as other experts on dyslexia will be sharing their stories at a luncheon next door in Rayburn Room 2325 immediately following this hearing, and we welcome all of you all to attend that luncheon.

For most people, dyslexia is a disability. But if we change the way we approach it, we can turn disability into possibility and give millions of individuals a brighter and more productive future.

[The prepared statement of Mr. Smith follows:]

PREPARED STATEMENT OF CHAIRMAN LAMAR S. SMITH

Welcome everyone to today's hearing on the Science of Dyslexia. One out of every five people struggle with dyslexia in its various forms. In fact, it is the most common reading disability in America. Yet many Americans remain undiagnosed, untreated, and silently struggle with school or work.

People with dyslexia think in a way that others do not. But typically in our school systems today there is not recognition, early detection, or enough teachers who are trained to spot symptoms of dyslexia early enough to get the students the intervention they need.

That is why we have recently seen grass roots groups, like Decoding Dyslexia, form nationwide and more specialized schools started to fill the void. Unfortunately, not everyone has access to these types of schools and the learning strategies they instill in their students to help them become successful.

I hope today's hearing will serve two purposes. First, contribute to our understanding as policy-makers about the neuroscience of dyslexia. And secondly, build awareness of dyslexia's effect on those of all ages if we fail to diagnose it.

Some may ask why the Science Committee chooses to tackle the issue of dyslexia. My response is simple: many scientists, innovators and other outside-the-box thinkers are dyslexic, such as Albert Einstein, Leonardo da Vinci, and Galileo, to name a few.

Many who have dyslexia have used their unique outlook on the world to their advantage. Filmmakers, actors and entertainers such as Steven Spielberg, Henry Winkler, and Jay Leno used their gift to create one-of-a-kind entertainment for us all to enjoy.

In modern times, Dr. Carol Greider of the Johns Hopkins School of Medicine, who won the Nobel Prize in 2009, has dyslexia. John Chambers, the long-time CEO of Cisco Systems, also has dyslexia. In a recent interview, Chambers spoke about his struggles with dyslexia, saying:

> "It would surprise you how many government and business leaders [have] dyslexia. Some people view it as a weakness and maybe it is. Because of my weakness I've learned other ways to accomplish the same goal with faster speed. So in math, I can do equations faster by eliminating the wrong answers quicker than I can get the right answer. It's one of the reasons I talk to young people with dyslexia pretty regularly. You have to have role models."

We need to unleash the intelligence of people with dyslexia, like Einstein, da Vinci, Carol Greider, and John Chambers. We cannot afford for young, talented students not to reach their potential.

I am glad to see the National Science Foundation fund studies in how astrophysicists with dyslexia view the universe differently due to the visual-spatial skills common in dyslexics. In fact, Matt Mountain, the lead astronomer and director of the Hubble Space Telescope Science Institute, has dyslexia.

Also, the National Institutes of Health is studying the neuroscience of dyslexia, including the work of our witnesses, Dr. Sally Shaywitz and Dr. Guinevere Eden, as well as funding studies on how dyslexic students can best learn.

Beyond the research, we will hear from someone with dyslexia, the parent of a dyslexic student, and an educator for those with learning disabilities like dyslexia.

I have a personal connection with dyslexia since my niece is dyslexic. And a favorite, young 10 year old friend named Leighton, who has dyslexia, has been with me on a Texas ranch. He may be challenged by language arts but he makes up for it with perfect eyesight and exceptional accuracy with his bb gun. And you don't want to compete with him playing Minecraft on his Ipad.

Over 80 members of Congress have joined the bipartisan Congressional Dyslexia Caucus co-chaired by Rep. Bill Cassidy and Science Committee member Julia Brownley. I thank them both for their work in helping educate the public about dyslexia and for advocating policies that support those individuals who have dyslexia.

I also want to acknowledge one of my constituents, Robbi Cooper, who traveled from Austin, Texas, to be here today. She has shared many stories with me about her son Ben who has dyslexia.

More parents as well as other experts on dyslexia will be sharing their stories at a luncheon next door in Rayburn room 2325 immediately following this hearing. All are welcome to attend.

For most people, dyslexia is a disability. But if we change the way we approach it, we can turn disability into possibility—and give millions of individuals a brighter and more productive future.

Chairman SMITH. That concludes my remarks, and the gentlewoman from Texas, the Ranking Member of the Committee, is recognized for hers.

Ms. JOHNSON. Thank you very much, Mr. Chairman, and thank you for holding this hearing, and I want to thank the two Co-

Chairs of the Congressional Dyslexia Caucus, Representative Brownley and Representative Cassidy, for being here today, and I look forward to your testimony.

I have known a number of people who have dyslexia. None of them I have found to be dumb. Even though dyslexia is a lifelong condition, with proper diagnosis and instruction, individuals with dyslexia can succeed in school and go on to have successful careers. We have an example of someone with dyslexia who has a successful career with us here today, and I look forward to hearing his story, Mr. Brooks.

As the Science, Space, and Technology Committee, we oversee agencies that fund research from the very basic to development and deployment across nearly the entire portfolio of the Federal R&D. We don't directly oversee the lead agency for dyslexia research, which is NIH. However, we have the important responsibility for oversight of the National Science Foundation, which supports fundamental research across a number of scientific fields that provide a foundation for dyslexia research. The National Science Foundation, as a leader in educational research, also funds learning science directly and indirectly related to dyslexia.

A significant amount of the National Science Foundation research relevant to dyslexia is funded out of the Social, Behavioral, and Economic Sciences Directorate and the Education and Human Resources Directorate. This is why I have fought hard against efforts in this Committee to slash funding for these important NSF directorates, which fund valuable research that turns out to have broader, and often unanticipated, applications to other high-priority research as we are seeing here today.

Now, I know that the Chairman has asked me not to mention this because he didn't want this to be a partisan hearing. It is not partisan with me, it is reality. Additionally, research funding by Biological Sciences Directorate also contributes to funding foundational knowledge about the neuroscience behind dyslexia. The National Science Foundation Science of Learning Centers Program supports interdisciplinary centers that advance learning research, and I look forward to hearing from Dr. Guinevere Eden about the Gallaudet University Center, of which she is a part. That center focuses on visual information learning research.

Additionally, I am interested in hearing from Dr. Peter Eden about the National Science Foundation-funded research that is being conducted at Landmark College. This research is investigating important questions, including how students with learning disabilities best learn STEM, and how online educational environments could be adapted for students' learning disabilities. This research has the potential to improve educational outcomes of students with learning disabilities, including dyslexia, and perhaps for all students.

This learning demonstrates how important it is to fund our research agencies at appropriate levels. We have learned so much about dyslexia, but have so much more to learn. Without learning and—without funding research into this area, including the foundational research that underlies the more applied work, we will not discover the biological basis for dyslexia, we will not create the next generation of treatments for dyslexia, and we will not determine the educational environments and techniques that are best for individuals with dyslexia.

I want to thank our witnesses. I am sorry that Ms. Brownley had to change her testimony but I hope we will keep in mind as you make your statements that the only way we reach these goals is to fund the research.

Thank you, and I yield back, Mr. Chairman.

[The prepared statement of Ms. Johnson follows:]

PREPARED STATEMENT OF RANKING MEMBER EDDIE BERNICE JOHNSON

Thank you, Mr. Chairman for holding this hearing.

And I want to thank the two co-chairs of the Congressional Dyslexia Caucus, Representatives Brownley and Cassidy, for being here today. I look forward to hearing your testimony.

I have known a number of people who have dyslexia. Even though dyslexia is a lifelong condition, with proper diagnosis and instruction, individuals with dyslexia can succeed in school and go on to have successful careers. We have an example of someone with dyslexia who has a successful career here with us today. I look forward to hearing your story, Mr. Brooks.

As the Science, Space, and Technology Committee, we oversee agencies that fund research from the very basic to development and deployment across nearly the entire portfolio of federal R&D. We don't directly oversee the lead agency for dyslexia research, which is NIH. However, we do have the important responsibility for oversight of the National Science Foundation, which supports fundamental research across a number of scientific fields that provide a foundation for dyslexia research. The National Science Foundation, as a leader in educational research, also funds learning science directly and indirectly related to dyslexia.

A significant amount of the NSF research relevant to dyslexia is funded out of the Social, Behavioral, and Economic Sciences Directorate and the Education and Human Resources Directorate. That is why I have fought efforts in this Committee to slash funding for these important NSF Directorates, which fund valuable research that turns out to have broader, and often unanticipated, applications to other high-priority research—as we are seeing here today. Additionally, research funded by the Biological Sciences Directorate also contributes to foundational knowledge about the neuroscience behind dyslexia.

The NSF's Science of Learning Centers Program supports interdisciplinary centers that advance learning research. I look forward to hearing from Dr. Guinevere Eden about the Gallaudet University Center of which she is a part; that center focuses on visual information learning research.

Additionally, I am interested in hearing from Dr. Peter Eden about the NSF-funded research that is being conducted at Landmark College. This research is investigating important questions, including how students with learning disabilities best learn STEM, and how online educational environments could be adapted for students with learning disabilities. This research has the potential to improve educational outcomes for students with learning disabilities, including dyslexia, and perhaps for all students.

This hearing demonstrates how important it is to fund our research agencies at appropriate levels. We have learned so much about dyslexia, but have much more to learn. Without funding research into this area, including the foundational research that underlies the more applied work, we will not discover the biological basis for dyslexia. We will not create the next generation of treatments for dyslexia. And we will not determine the educational environments and techniques that are best for individuals with dyslexia.

I want to thank the witnesses on both panels for being here today. I look forward to their testimony and the Q&A.

Thank you, Mr. Chairman and I yield back the balance of my time.

Chairman SMITH. I think it is safe to say that everyone in this room favors more money for research in regard to dyslexia. It is clearly in the national interest for us to do so.

I would like to recognize our two witnesses, and both are Members of Congress, on our first panel, and our first witness is the Hon. Bill Cassidy, who represents Louisiana's 6th Congressional

District. Prior to his election to Congress, Representative Cassidy served in the Louisiana State Senate. He also served as an Associate Professor of Medicine at Louisiana State University and taught doctors-in-training at the Earl Long Hospital in Baton Rouge. Representative Cassidy serves on the Energy and Commerce Committee and is a Co-Chair of the bipartisan Congressional Dyslexia Caucus. Representative Cassidy received his undergraduate and medical degrees from Louisiana State University.

Our next witness is a Science Committee Member herself, Representative Julie Brownley, who represents California's 26th Congressional District. Representative Brownley began her career in public service as a school board member in California. She also served in the California State Legislature. Prior to her career in public service, Representative Brownley worked in marketing in private industry. In addition to her being a Member of the Science Committee, she also serves on the Veterans Affairs Committee and is a Co-Chair of the bipartisan Congressional Dyslexia Caucus. Representative Brownley graduated from George Washington University and earned an MBA from American University.

We welcome both of you, of course, for many reasons, and Dr. Cassidy, if you will begin?

TESTIMONY OF THE HON. BILL CASSIDY, MEMBER, U.S. HOUSE OF REPRESENTATIVES

Dr. CASSIDY. Thank you, Chairman Smith and Ranking Member Johnson, for inviting me to speak, for us to be here for this bipartisan meeting to bring attention to the science of dyslexia.

As you said, Mr. Chairman, 20 percent of the U.S. population is dyslexic, dyslexia affects as many as 10 million children, boys, girls, all ethnic, socioeconomic and geographic regions of our country.

It is an important issue to me as a parent and as a Congressman. A couple of years ago, my youngest daughter was diagnosed with dyslexia. I shouldn't be upset because it is only a diagnosis. On the other hand, the struggles we had to do to have my daughter accommodated were something that I wouldn't care for other parents. So again, I thank you.

Now, prompted by concerns about my daughter and my constituents' children, my wife and I set out to learn as much as we could about dyslexia, and we were amazed at how much is known and yet not incorporated into public policy. Once more, thank you for highlighting the science.

Debra Stark is here, and you know, Debra has a child with Pete Stark with dyslexia, and my wife came to a conference that she sponsored here with the Shaywitzes, and Pete and are I are good husbands. We did exactly what our wives told us to do. We co-founded the Congressional Dyslexia Caucus. Pete has left Congress and so now Julia is also the Co-Chair, and I appreciate that. And the purpose of the caucus is to educate other Members of Congress and advance policies, to break down barriers faced by dyslexics.

Now, I firmly believe by raising awareness of dyslexia we can change the way we educate our children and assist millions of children to get onto the pathway of success. Now, part of this is the resolution that the two of us introduced. It urges the House of Rep-

12

resentatives to call upon schools along with state and local educational agencies to address the implications that dyslexia has on a student's education, and we now have over 100 Members of Congress on this resolution.

Now, dyslexia robs an individual of the ability to read quickly and automatically and to retrieve spoken words easily but it does not dampen their creativity and ingenuity. As you mentioned, some of the best and brightest among us are dyslexics. A few examples: Charles Schwab and the late Steve Jobs.

Now, if dyslexia is identified in elementary school and appropriate resources are made available, America can have more teachers, scientists, entrepreneurs, Charles Schwabs and Steve Jobs.

Now, science shows the reading pathway in the brain of those who are dyslexic is different. MRIs show a specific disruption of the reading system. Those affected need an evidence-based curriculum addressing this reading disruption. Now, unless accommodations are made, curriculums and trained teachers are applied that correspond to the science of dyslexia, children will languish in the classroom. A one-size-fits-all approach does not meet the needs of a dyslexic.

Now, for those with money, there are excellent schools in areas of the country where your child will learn to read and have all the opportunities that reading allows. But if a family cannot afford a $10,000 to $50,000 annual tuition, the option is often a traditional public school in which dyslexics are mainstreamed, which is to say, they don't have this particular curriculum, will not likely receive the remediation they need and have the future that the inability to read predicts.

So I applaud schools and educators who have embraced science by providing students with the proper educational environment and curriculum that allows them to thrive personally and academically. There are schools in Louisiana, like the Louisiana Key Academy— we will hear from a parent whose child attends that school—in Baton Rouge and the Max Charter School in Thibodaux, Louisiana, that specialize in teaching dyslexic students. But these schools are too few and far between. We need more schools to embrace and replicate this model so that students can achieve their fullest potential.

You mentioned there was a festive attitude. There is a festive attitude. If you are a parent or a child who has had this condition and no one ever seemed to acknowledge it, the science was hocus pocus, they didn't accept it, even though the science is real, you are celebrating that this Committee is elevating the status of that science. So I believe that we can come together on behalf of the children we love and the Nation we serve and work in a bipartisan and bicameral capacity. Greater strides need to be made in ensuring that every dyslexic child and adult has a chance to read, to learn, to demonstrate and to realize his or her full potential.

So thank you again for holding this hearing and giving the science behind dyslexia the attention it deserves. Hopefully, our work with the resolution, the caucus, and this hearing will have a positive impact upon society and everyone striving to learn with dyslexia.

[The prepared statement of Dr. Cassidy follows:]

Congressman Bill Cassidy, M.D.
Testimony- The Science of Dyslexia Hearing
Committee on Science, Space & Technology
September 18, 2014

Thank you Chairman Smith and Ranking Member Johnson for inviting me to speak and hosting this bipartisan hearing bringing attention to the science of dyslexia.

Twenty percent of the United States population is dyslexic. This disability affects as many as 10 million children across the country — boys and girls from all ethnic, socioeconomic and geographic regions of our country.

It is an important issue for me, both as a parent and as a Congressman.

A couple of years ago, my youngest daughter was diagnosed with dyslexia. Prompted by concerns about my daughter and my constituents' children, I set out to learn as much as I could about dyslexia and was amazed at how much is known and yet, far too often, not incorporated into public policy and education.

As a result, I co-founded the bipartisan Congressional Dyslexia Caucus and chair it with Congresswoman Brownley. The purpose of the caucus is to educate other Members of Congress and advance policies to break down barriers faced by dyslexics. I firmly believe that by raising awareness of dyslexia we can change the way we educate our children and assist millions of children to get on the path to success.

Part of this effort is a resolution I introduced along with Congresswoman Brownley. Our resolution urges the House of Representatives to call on schools along with state and local educational agencies to address the implications that dyslexia has on students. We now have over 100 Members of Congress cosponsoring this resolution.

Dyslexia robs individuals of their ability to read quickly and automatically and to retrieve spoken words easily but it does not dampen their creativity and ingenuity. We know that many with dyslexia are among our brightest and most successful. A few examples include entrepreneurs such as Charles Schwab and the late Steve Jobs.

If dyslexia is identified in elementary school and the appropriate resources are given to these children, America can produce more teachers, more scientists and more entrepreneurs.

Science shows the reading pathway in the brain of those who are dyslexic is different. MRI's show a specific disruption of the reading system. Those affected need an evidence based curriculum to address this reading disruption. Unless accommodations are made; curriculums and trained teachers are applied that correspond to the science of dyslexia, children will languish in the classroom. A one-size-fits-all approach will not work.

For those with money, there are excellent schools in some areas of the country where your child will learn to read and have all the opportunities reading allows. If a family cannot afford a $10,000 to $50,000 annual tuition, the option is typically a traditional public school in which dyslexics are "mainstreamed", which is to say, they likely will not receive the remediation they need.

So I applaud schools and educators who have embraced science by providing students with the proper educational environment and curriculum that will enable them to thrive personally and academically.

There are schools in Louisiana, like Louisiana Key Academy in Baton Rouge and the Max Charter School in Thibodaux, that specialize in teaching dyslexic students. But these schools are too few and far between. We need more schools to embrace and replicate this model so students can reach their full potential.

I believe we can come together on behalf of the children we love and the nation we serve and work in a bipartisan and bicameral capacity. Greater strides need to be made in ensuring that every dyslexic child and adult has a chance to read, to learn, to demonstrate, and to realize his or her full potential.

Thank you again for holding this hearing and giving the science behind dyslexia the attention it deserves. Hopefully, our work with the resolution, the caucus, and this hearing will have a positive impact on society and everyone striving to learn with dyslexia.

Thank you.

Chairman SMITH. Thank you, Dr. Cassidy, and Representative Brownley.

TESTIMONY OF THE HON. JULIA BROWNLEY, MEMBER, U.S. HOUSE OF REPRESENTATIVES

Ms. BROWNLEY. Thank you, and good morning to you all, and I too want to begin by thanking Chairman Smith and Ranking Member Johnson for your leadership, and for inviting me to testify this morning about dyslexia. I also want to thank the witnesses who have come here to discuss their research and experiences with dyslexia.

When my daughter Hannah struggled to learn to read in third grade, then fourth grade and fifth grade, like any parent I was completely panicked about what to do next. It was really Hannah's dyslexia, and my own learning to navigate the school system, where I frankly witnessed the good, the bad and the ugly that surely led me to public service, first, as you mentioned, as a school board member, then as a member of the California State Legislature, and now in Congress.

This spring, Hannah received her master's degree in international studies and is now overseas saving the world with a NGO in Kenya, Africa. She speaks three languages and, not surprisingly, she still misspells in each one of them. I could not be prouder of her. But for every success story like my daughter Hannah, there are countless others who do not succeed.

Learning disabilities like dyslexia and attention-related disorders affect as many as one in five children in the United States. According to the National Center for Learning Disabilities, nearly half of secondary students with learning disabilities like dyslexia perform more than three grade levels below their enrolled grade in essential academic skills—45 percent in reading, 44 percent in math. Twenty percent of students with a learning disability drop out of high school, compared to just eight percent of students in the general population. That is millions of American children who aren't reaching their full potential.

However, advancements in cognitive science are teaching us much more about how the brain develops and how children learn. My hope is that today's hearing will inform lawmakers about how to better translate groundbreaking research to innovative education policy that will make a difference in the lives of millions of Americans with dyslexia. Our education system needs to do a better job training teachers to recognize and effectively educate students with dyslexia. We need to provide our schools with the resources to incorporate assistive technologies, such as audiobooks and speech-to-text interfaces, in the classroom, as well as support services to ensure every child has an equal opportunity to succeed.

The Federal Government also needs to meet its obligations to our schools. For decades, Congress has failed to meet its 40 percent financial commitment for special education costs under the Individuals with Disabilities Education Act, placing a heavier and heavier burden on states and local schools. Congress also needs to increase its investment in scientific research on dyslexia.

I am sure many of my colleagues have heard from frustrated parents in their districts who are very concerned about their children with dyslexia—will they be allowed to take advanced placement courses, or pursue a passion like music or science? Will they get reasonable accommodations on state tests and college entrance exams? What will happen when their children graduate from high school and make the transition to college?

One of our Committee's most important missions is creating a 21st century workforce of engineers, scientists, and STEM professionals. To accomplish that goal, we need to make sure every student has the support they need from their educators, parents, their communities to succeed. Students with dyslexia are smart and capable and perhaps uniquely qualified because of their out-of-the-box way of attacking problems and processing information, but misconceptions about dyslexia too often result in a focus on a disability rather than an ability. Today's panelists will demonstrate that this community of young people have extraordinary strengths, and that ignoring dyslexics costs us all.

If you think that more should be done to address dyslexia, as my Dyslexia Caucus Co-Chair has already mentioned, please cosponsor our Dyslexia Caucus bipartisan resolution, H.R. 456. We already have 107 cosponsors, and we would welcome your support.

And with that, I thank you, Mr. Chairman, and I yield back the balance of my time.

[The prepared statement of Ms. Brownley follows:]

The Honorable Julia Brownley
Testimony before the House Science, Space, and Technology Committee
Hearing on "The Science of Dyslexia"
September 18, 2014

Good morning.

I want to begin by thanking Chairman Smith and Ranking Member Johnson for their leadership, and for inviting me to testify about dyslexia.

I also want to thank the witnesses who have come here to discuss their research and experiences with dyslexia.

When my daughter Hannah struggled to learn to read, like any parent I was completely panicked about what to do next. It was Hannah's dyslexia, and learning to navigate the school system, where I witnessed the good, the bad, and the ugly, that led me to public service: first as a school board member, then in the State Legislature, and now in Congress. This spring, Hannah received her Master's degree in International Studies, and is now overseas saving the world with a NGO in Kenya, Africa. She speaks three languages, and she still misspells in all of them. I could not be prouder of her. But for every success story like Hannah, there are countless others who do not succeed.

Learning disabilities, like dyslexia, and attention-related disorders affect as many as 1 in 5 children in the United States. According to the National Center for Learning Disabilities, nearly half of secondary students with learning disabilities like dyslexia perform more than three grade levels below their enrolled grade in essential academic skills (45% in reading, 44% in math). 20% of students with a learning disability drop out of high school, compared to just 8% of students in the general population. That's millions of American children who aren't reaching their full potential.

However, advancements in cognitive science are teaching much more about how the brain develops, and how children learn. My hope is that today's hearing will inform lawmakers about how to better translate groundbreaking research to innovative education policy that will make a difference in the lives of millions of Americans with dyslexia.

Our education system needs to do a better job training teachers to recognize and effectively educate students with dyslexia. We need to provide our schools with the resources to incorporate assistive technologies, such as audiobooks and speech to text interfaces, in the classroom, as well as support services to ensure every child has an equal opportunity to succeed.

The federal government also needs to meet its financial obligations to our schools.

For decades, Congress has failed to meet its 40% financial commitment for special education costs under the Individuals with Disabilities Education Act, placing a heavier burden on states and local schools.

Congress also needs to increase its investment in scientific research on dyslexia, and other areas.

I'm sure many of my colleagues have heard from frustrated parents in their districts who are very concerned about their children with dyslexia: will they be allowed to take advanced courses, or pursue a passion like music or science? Will they get reasonable accommodations on state tests and college entrance exams? What will happen when their child graduates high school and makes the transition to college, where resources for students with learning disabilities can even more difficult to find than in elementary and secondary schools? Too often, misconceptions about dyslexia or learning disabilities result in a focus on a disability rather than an individual child's ability,

One of our Committee's most important missions is creating a 21st century workforce of engineers, scientists, and STEM professionals. To accomplish that goal, we need to make sure every student has the support they need from their educators, parents, and the community, to succeed.

Students with dyslexia are smart and capable and perhaps uniquely qualified because of their out-of-the-box way of attacking problems and processing information, but misconceptions about dyslexia too often result in a focus on a disability rather than ability. Today's panelists will demonstrate that this community has extraordinary strengths, and that ignoring dyslexics costs us all.

If you think that more should be done to address dyslexia, I highly encourage you to cosponsor the Congressional Dyslexia Caucus' bipartisan resolution, H.Res. 456. We already have 107 cosponsors, and we would welcome your support.

Chairman SMITH. Thank you, Representative Brownley, and I hope every Member of the Science Committee will cosponsor that resolution and we will urge them to do so.

Thank you both for your testimony. We will have some questions that will be submitted perhaps by Members that will be in writing and perhaps you can get back to us when they submit those questions.

We will now move to our next panel. Of course, Ms. Brownley is a Member of this Committee, and Dr. Cassidy, if you would like to join us up here, you are welcome to do that as well, and again, thank you both.

Let me introduce our second panel of witnesses, and our first witness is Dr. Sally Shaywitz, an Audrey G. Ratner Professor in Learning Development at the Yale University School of Medicine and Co-Director of the Yale Center for Dyslexia and Creativity. Dr. Shaywitz has authored more than 200 scientific articles and books, and together with her husband, Dr. Bennett Shaywitz, she is the originator of the Sea of Strength's model of dyslexia. Dr. Shaywitz is also an elected Member of the Institute of Medicine of the National Academy of Sciences. She received her bachelor's degree from the City University and her M.D. from Albert Einstein College of Medicine.

Our second witness is Max Brooks, a New York Times bestselling author and actor and a screenwriter. He is known for his popular books, the Zombie Survival Guide and World War Z as well as for his roles in television shows such as Rosanne, Pacific Blue and Seventh Heaven. He also worked as a member of the writing team of Saturday Night Live. Mr. Brooks received his bachelor's degree from American University.

Our next witness is Stacy Antie, a parent advocate for educational interventions for dyslexia and other language-based learning disabilities within our schools. Ms. Antie is the mother of a 9-year-old boy who was diagnosed with dyslexia and who currently attends Louisiana Key Academy. She has been spreading her son's story for the past year to help other families in similar situations. She is a lifelong Louisiana native.

Our next witness is Dr. Peter Eden, the President of Landmark College. Previously, Dr. Eden served as the Dean of Arts and Sciences and Professor of Biotechnology at Endicott College, Associate Professor and Chair of the Science Department at Marywood University and Research Fellow at the Jackson Laboratory. He has also worked as a molecular biologist and Research Project Director at Biomeasure Inc. Dr. Eden has published more than 20 scientific articles and has received funding for his research from the National Institutes of Health and the National Science Foundation. Dr. Eden received his bachelor's degree from the University of Massachusetts and his Ph.D. from the University of New Hampshire.

Our final witness, Dr. Guinevere Eden, is an Associate Professor in the Department of Pediatrics and the Department of Psychology at Georgetown University and an Adjunct in the Department of Pediatrics at George Washington University. She also serves as an advisor of Great Schools Inc. and directs the Center for the Study of Learning. Dr. Eden previously served as President of the Board of the International Dyslexia Association where she now serves as the Director, and on the editorial boards of the scientific journals Annals of Dyslexia and Human Brain Mapping. Dr. Eden received her bachelor's degree from University College London and her Ph.D. from Oxford University.

We welcome you all, and Dr. Shaywitz, if you will lead us off?

TESTIMONY OF DR. SALLY SHAYWITZ, AUDREY G. RATNER PROFESSOR IN LEARNING DEVELOPMENT, YALE UNIVERSITY SCHOOL OF MEDICINE AND CO-DIRECTOR, YALE CENTER FOR DYSLEXIA AND CREATIVITY, YALE UNIVERSITY

Dr. SHAYWITZ. Good morning, Chairman Smith, Ranking Member Johnson and other Committee Members. Thank you for the opportunity to speak with you about the science of dyslexia and share with you the tremendous scientific progress that has been made in dyslexia.

In dyslexia, there is an abundance of high-quality scientific knowledges so that we do not have a knowledge gap but an action gap. It is our hope hearing the depth and breadth of the scientific knowledge of dyslexia will alert policymakers to act and to act with a sense of urgency.

Resolution 456 submitted by Representative Bill Cassidy of Louisiana and Representative Brownley provides an up-to-date universal definition of dyslexia, incorporating scientific advances and understanding of dyslexia, especially its unexpected nature, and represents a landmark in aligning science and education.

Dyslexia represents 80 to 90 percent of all learning disabilities and differs markedly from all others in that dyslexia is very specific and scientifically validated. We know its prevalence, cognitive and neurobiologic origins, symptoms and effective interventions. Learning disabilities is a general term, referring to a range of difficulties, most of which have not yet been delineated or scientifically validated.

Data from sample surveys indicate dyslexia is very common, affecting, as you have heard, one out of five. Yes, you heard correctly. It is not the prevalence often quoted by schools. Why? The why is the reason we are here today. Schools far too often fail to acknowledge, much less identify, students who are dyslexic. Initial descriptions of dyslexia as an unexpected difficulty in reading are empirically validated as demonstrated in this slide. Slide, please.

In typical readers, IQ and reading track together over time. As you can see on the left, in typical readers' IQ and reading go together. In contrast, in dyslexic readers, reading and intelligence are not linked so a child can have a very high IQ, and unexpectedly read at a much lower level. Dyslexia reflects a difficulty within the language system, more specifically, the phonologic component of language. It is not seeing words backwards.

Dyslexia is a paradox. The same slow reader is often a very fast and able thinker, giving rise to our conceptual model of dyslexia as a weakness in getting to the sounds of spoken words surrounded by a sea of strengths in higher-level thinking processes such as reasoning and problem-solving. Reflecting this paradox are many eminent dyslexics, as you have heard, Charles Schwab, David Boies and Dr. Toby Cosgrove, who is CEO of the Cleveland Clinic and a cardiac surgeon.

On the other side of the coin, though, are many who are not identified and do not receive evidence-based instruction, struggle to read and come to see themselves as a failure. Converging evidence from our own and other laboratories has identified a neural signature for dyslexia. Slide, please.

[Slide.]

That is an inefficient functioning of those posterior left hemisphere language and reading systems.

Recent data—slide—from our laboratory indicate that the gap between typical and dyslexic readers is already present by first grade, and persists, a very clear message: we have to get to these children very early and not wait.

Yes, dyslexic children can learn to read and must be taught to read. It is imperative that teachers and parents learn about the powerful science of dyslexia, know how to identify dyslexia early on, and provide evidence-based methods to teach dyslexic children to read. We must not give up on teaching reading and limit a child's future options. Education must and can be aligned with science.

We must ensure that scientific knowledge is translated into policy and practice and that ignorance and injustice do not prevail. We know better and we must act better.

I cannot look into the face of one more child who has lost faith in himself and the world. I cannot look into the face of a child's father who is desperately trying to hold back tears. I cannot hear once again about how a school told a mother, we do not believe in dyslexia.

As shown in this next slide, an iceberg is 90 percent underwater with only ten percent visible. Similarly, in dyslexia, we hear about the ten percent like Max Brooks who have made it. Let us not give up on the invisible 90 percent still underwater asking, indeed begging, to be helped.

I am optimistic, though. Once this Committee is aware of the strong science of dyslexia, educators will want to align their practices and policies with 21st century science. It is good for the children, for their families and for our Nation.

Thank you.

[The prepared statement of Dr. Shaywitz follows:]

Testimony Before the

Committee on Science, Space, and Technology

United States House of Representatives

Statement of

Sally E. Shaywitz, M.D.

The Audrey G. Ratner Professor in Learning Development

Co-Director, Yale Center for Dyslexia & Creativity

Yale University School of Medicine

Thursday September 18, 2014

Good morning Chairman Smith, Ranking member Johnson, and other committee members. Thank you for the opportunity to speak with you about the science of dyslexia and share with you the tremendous scientific progress that has been made in dyslexia.

My name is Sally Shaywitz and I am a physician-scientist, The Audrey G. Ratner Professor in Learning Development, and Co-Director of the Yale Center for Dyslexia & Creativity at the Yale University School of Medicine. I am a member of the National Academy of Medicine of the National Academy of Sciences, and have served on the Advisory Council of the National Institute of Neurological Diseases and Stroke (NINDS), the National Research Council Committee on Women in Science and Engineering, co-chaired the National Research Council Committee on Gender Differences in the Careers of Science, Engineering and Mathematics Faculty and have served on the Congressionally-mandated National Reading Panel and the Committee to Prevent Reading Difficulties in Young Children of the National Research Council. I am also the recipient of an Honorary Doctor of Science degree from Williams College. I am presenting these research data on behalf of myself and Bennett Shaywitz, MD, the Charles and Helen Schwab Professor in Dyslexia and Learning Development and Co-Director of the Yale Center for Dyslexia & Creativity who leads the Center's research program.

I speak to you as a physician-scientist. As a physician, I have all too many memories of sitting by an ailing child's bedside, wishing so desperately that we had the knowledge to help that child. As a physician I know the power of science and how once new knowledge becomes available we act quickly, indeed, race to put that knowledge to good use. We want to close that knowledge gap and improve the lives of the affected children. When I sat on the Advisory Council of the National Institute of Neurological

Disorders and Stroke, we constantly asked ourselves: how have we benefited mankind, how has our research improved the well-being of children and adults.

As you will hear, in dyslexia, science has moved forward at a rapid pace so that we now possess the data to reliably define dyslexia, to know it's prevalence, it's cognitive basis, it's symptoms and remarkably, where it lives in the brain and evidence-based interventions which can turn a sad, struggling child into not only a good reader, but one who sees herself as a student with self-esteem and a fulfilling future.

In dyslexia: an action gap

So what's the problem? The good news is that our problem is a solvable one. Of course, we are always seeking new knowledge. In dyslexia there is sufficient high quality scientific knowledge to help and to turn around the lives of so many struggling children. In dyslexia, remarkably in America, in the year 2014, we have not a knowledge gap but an action gap. We have the knowledge but it is not being put into policy and practice and far too many children and adults, too, are suffering needlessly. There is an epidemic of reading failure that we have the scientific evidence to treat effectively and we are not acknowledging or implementing it. It is our hope that hearing the depth and extent of the scientific knowledge of dyslexia will alert policy makers to act and to act with a sense of urgency.

The really good news: Science is there for those who are dyslexic.

Science informing education

Resolution 456 submitted by Rep. Bill Cassidy, on behalf of the Bipartisan Dyslexia Caucus which he co-founded and currently Co-Chairs, provides the most up-to-date, universal, scientifically valid definition of dyslexia incorporating scientific advances in understanding dyslexia, especially, its unexpected nature, and represents a landmark in aligning science and education. Furthermore, Resolution 456 notes the "diagnosis of dyslexia is critical and must lead to focused, evidence-based interventions, and necessary accommodations..."

Dyslexia is specific; learning disabilities are general

Dyslexia is the most common and most carefully studied of the learning disabilities, affecting 80% to 90% of all individuals identified as learning disabled. Of the learning disabilities, dyslexia is also the best characterized and the oldest. In fact, the first description of dyslexia preceded the first mention of learning disability by over sixty years – dyslexia was first reported by British physician, Dr. Pringle Morgan, in 1896, describing Percy F., "He has always been a bright and intelligent boy, quick at games, and in no way inferior to others of his age. His great difficulty has been – and is now – his inability to learn to read." – a description that characterizes the boys and girls, men and women, I continue to see to this day. In contrast, the term learning disabilities was first used only in 1962.

Dyslexia differs markedly from all other learning disabilities. Dyslexia is very specific and scientifically validated: we know its prevalence, cognitive and neurobiological origins, symptoms, and effective, evidence-based interventions. Learning disabilities is a general term referring to a range of difficulties which have not yet been delineated or scientifically validated. Learning disabilities are comparable to what in medicine are referred to as 'infectious' diseases, while dyslexia is akin to being diagnosed with a strep throat – a highly specific disorder in which the causative agent and evidence-based treatment are both known and validated.

Epidemiology of dyslexia: prevalence

Scientific studies in a range of disciplines provide epidemiologic, cognitive and neurobiological data to characterize dyslexia. Epidemiologic data from sample surveys in which *each* individual is assessed indicate that dyslexia is highly prevalent, affecting one in five, yes you read this correctly. It is not the stated prevalence often quoted. Why? The why is the reason we are here today – schools far too often fail to acknowledge, much less identify, students who are dyslexic. Consequently, schools will report low, but incorrect number of students affected. *If they are not identified, they cannot be counted.*

Many believe that even this 1 in 5 estimate may be too low. For example, data from the 2013 National Assessment of Educational Progress (NAEP, the Nation's Report Card) indicate that 2 in 3 students in 4th or 8th grade are not proficient readers. Among some groups of students the numbers are far worse. The NAEP data show that 4 in 5 African-American, Latino and Native American students are not proficient readers. Many would consider this to be an out-of-control epidemic of reading failure, and considering its negative consequences, a national crisis demanding action. Longitudinal studies, prospective and retrospective, indicate dyslexia is a persistent, chronic condition; it does not represent a "developmental lag."

Sample surveys in which every subject has been individually assessed show relatively equal numbers of males and females affected. Studies based on school-based identification show a high male prevalence with accompanying data indicating that the often disruptive behaviors of the boys in the classroom play a strong role in bringing them to the attention of their teacher with subsequent referral. Girls who may be struggling readers, but who are sitting quietly in their seats, far too often fail to be identified.

Dyslexia has no known boundaries, it is universal, affecting virtually all geographic areas, and both alphabetic and logographic languages. For example, my book, *Overcoming Dyslexia,* (Knopf) has been translated, as expected, into alphabetic languages (Portuguese, Dutch, Croatian, etc.) but also, a surprise to me, logographic scripts including Japanese and Korean, and most recently, Chinese. In addition, dyslexia occurs in every ethnic, race and socio-economic class.

Unexpected nature of dyslexia

Dr. Morgan's initial description of dyslexia over 100 years ago as an *unexpected* difficulty in reading has now been validated by empiric evidence. Our research group found that in typical readers, IQ and reading are dynamically linked, they track together over time and influence each other. In contrast, in dyslexic readers, reading and intelligence are not linked and develop more independently so that a child can have a very high IQ and, *unexpectedly*, read at a much lower level.

Cognitive basis of dyslexia

Dyslexia is a difficulty within the language system, more specifically, the phonological component of language – it is not seeing words backwards. Data from laboratories around the world now answer the question – why do otherwise bright and motivated children struggle or even fail to learn to read? Almost invariably, they have a phonologic deficit. To explain, converging evidence over the past several decades supports the phonological basis of dyslexia. Phonological refers to the smaller pieces of language that make up a spoken word. To understand the implications of this theory, we compare what we know about spoken compared to written language. Spoken language is natural and does not have to be taught - everyone speaks. Reading is artificial and must be taught. The key in learning to read is that the letters have to be linked to something that has inherent meaning – the sounds of spoken language. To read, the beginning reader must come to recognize that the letters and letter strings represent the sounds of spoken language. She has to develop the awareness that spoken words can be pulled apart into their basic elements, phonemes, and that the letters in a written word represent these sounds. Children and adults who are dyslexic struggle to pull apart the spoken word and, as a result, cannot isolate each sound and attach it to its letter. Results from large and well-studied populations of dyslexic children confirm that in young children as well as adolescents a deficit in phonology represents the most specific and robust correlate of dyslexia.

With the phonologic deficit recognized and validated, it is now possible to understand and to predict the symptoms emanating from this basic difficulty, which can be both observed and measured, resulting in an accurate diagnosis of dyslexia. Dyslexia is a language based difficulty and impacts spoken language, for example, word retrieval difficulties; reading, initially impacting reading accuracy and then reading fluency, the ability to read not only accurately, but also rapidly and automatically with good understanding. Not being able to read automatically, dyslexic readers must read what I refer to as 'manually,' requiring the output of large amounts of effort and consuming much of the individual's attention. A dyslexic reader lacks fluency meaning that he reads slowly and with great effort, although he may understand the content at a high level. Importantly, the dyslexic's vocabulary and comprehension may be quite high. Spelling is also problematic as is learning a foreign language.

The paradox of dyslexia

Dyslexia is a paradox, the same slow reader is often a very fast and able thinker – giving rise to our conceptual model of dyslexia as a weakness in getting to the sounds of spoken words surrounded by a sea of strengths in higher level thinking processes such as reasoning and problem solving. Reflecting this paradox are many eminent dyslexics - Charles Schwab, David Boies, and Dr. Toby Cosgrove. On the other side of the coin, are many who are not identified, do not receive evidence-based instruction, continue to struggle to read and see themselves as failures. Sadly, these boys and girls have no knowledge of what their difficulty is or that it even has a name, have no self-understanding, come to view themselves as dumb or stupid, see themselves as not meant for school, suffer low self-esteem, often drop out of school with a loss to themselves, to their families and to society.

Neurobiology of dyslexia

Converging evidence using functional magnetic resonance imaging (fMRI) from our own and laboratories around the world has identified three major neural systems for reading in the left hemisphere, one region, anterior, in Broca's area and two regions posterior, one in the parieto-temporal (or Wernicke's area), and another, in the occipito-temporal region, often referred to as the word form area. Furthermore, such fMRI studies indicate that in dyslexic readers, the posterior neural systems are functioning inefficiently, providing a *neural signature for dyslexia*. Critically, these posterior neural systems appear to be important in skilled, automatic reading and inefficient functioning in these neural systems suggest an explanation for the slow, effortful reading characterizing dyslexic readers. Recent studies of brain connectivity by us and others demonstrate that in dyslexic readers there is reduced connectivity to the posterior neural systems responsible for skilled, automatic reading.

Promising areas for research

In terms of promising areas for research, we believe it is important to better understand the relationship between reading and attention, the construct, and not ADHD, the disorder. Studies now in progress in our lab are examining the role of attention in reading, including probes of the relationship between those neural systems for attention and those for reading.

Reading gap already present by first grade and persists

Scientific knowledge, too, has delineated the progression of reading development. Reading growth is most rapid early on, during the first few years of school and then plateaus. Recent data from our laboratory indicate that the gap between typical and dyslexic readers is already present by first grade and persists. A very clear message: we have to get to these children very early and not wait.

National Reading Panel and teaching reading

Fortunately, thanks to Congressional action there is now strong evidence of what treatments are effective in teaching children to read. In 1998 Congress mandated the formation of a National Reading Panel to investigate the teaching of reading. I was proud to serve on the panel which produced the Report of the National Reading Panel. As a result, today it is no longer acceptable to use reading programs lacking scientific evidence of efficacy; instead it should be mandatory to use programs that are evidence-based, proven to be effective in the same way that medications must be tested and proven to be effective before they can be approved by the FDA. Our children deserve no less. And yet, today, this powerful information is not being used in schools, children are not learning to read and giving up, and not reaching their full potential. We have what amounts to an educational emergency in the US. Children are not learning to read with serious academic, economic and health consequences including, school drop-out, being half as likely to go on to college, significantly lower lifetime earning, significantly higher unemployment, higher rates of mental health issues such as often incapacitating anxiety, and, as reported in 2013, significantly higher mortality rates related to lack of a high school diploma. These harsh consequences harm not only the dyslexic individual but place our country at a competitive disadvantage.

Accommodations

Given that a student who is dyslexic has both a weakness and strengths, it is critical that for example, tests, both in school and on high stakes standardized examinations actually measure the student's ability and not his disability. The dyslexic student may learn to read fairly accurately but hardly ever with fluency; he remains a slow reader albeit a quick thinker. These dyslexic students may know the answer to a test question, but as a result of their slow reading never get to reach many questions or to finish the test, the student simply runs out of time. Or, she is so anxious about finishing the exam that she races through it and misses questions which, given the needed time, she would be able to answer correctly. Thus, it is critical that students who are dyslexic receive the accommodation of extra time; it is not a perk but a necessity if the result of the test is to reflect that student's knowledge. In adolescents and young adults applying for high-stakes standardized tests for college, graduate or professional schools, the Americans with Disability Amendment Act (ADAA) of 2008 is highly supportive of the need for accommodations for those with disabilities like dyslexia that impair a major life activity like reading. The ADAA regulations also state that students should receive accommodations even if they are doing well in school, it is not the outcome of their performance but rather what they have to do to achieve the outcome.

High school and college students with a history of childhood dyslexia often present a paradoxical picture; they may be similar to their unimpaired peers on measures of comprehension, but they continue to suffer from the phonologic deficit that makes reading less automatic, more effortful, and slow. Neurobiological data provide strong evidence for the necessity of extra time for readers with dyslexia. Functional MRI data demonstrate that in dyslexic readers the word-form area, the region supporting rapid reading, functions inefficiently. Readers compensate by developing anterior systems bilaterally and the right homolog of the left word-form area. Such compensation allows for more accurate reading, but it does not support fluent or rapid reading. For these readers with dyslexia, the provision of extra time is an essential accommodation, particularly on high stakes tests such as SAT, ACT and tests for professional schools such as LSAT, MCAT and GRE. The accommodation of extra time allows the student time to decode each word and to apply his unimpaired higher-order cognitive and linguistic skills to the surrounding context to get at the meaning of words that he cannot entirely or rapidly decode. While readers who are dyslexic improve greatly with additional time, providing additional time to non-dyslexic readers results in very minimal or no improvement in scores.

Although providing extra time for reading is by far the most common accommodation for people with dyslexia, other helpful accommodations include allowing the use of computers for writing essay answers on tests, access to recorded books and text to voice software. Other helpful accommodations include providing access to syllabi and lecture notes, tutors to "talk through" and review the content of reading material, alternatives to multiple-choice tests (e.g., reports or projects), waivers of high stakes oral exams, a separate, quiet room for taking tests, and a partial waiver of the foreign language requirement. Dyslexic students who have difficulty accessing the sound system of their primary language will, almost invariably, have difficulties learning a foreign language. Students with dyslexia most often have no difficulty with the mastery of high level courses, the problem lies in their lack of fluent, rapid reading so that it is the time necessary for them, as dyslexics, to read through the materials that is problematic. Many rigorous schools allow these students to take one course less during the school year and take this course during the summer. With such accommodations, many students with dyslexia are successfully completing studies in a range of disciplines, including science, law, medicine and education. It is accommodations such as these that are encouraging, and allowing, more students who are dyslexic to enter and to succeed in STEM fields.

Summary and Implications of the science of dyslexia

Yes, dyslexic children can learn to read and must be taught to read. It is imperative that teachers and parents learn about the powerful science of dyslexia, know how to identify dyslexia early on and to provide evidence-based methods to teach dyslexic children to read. We must not give up on teaching reading and limit a child's future options. Education must, and can be, aligned with science.

We must ensure that scientific knowledge is translated into policy and practice and that ignorance and injustice do not prevail. We know better, we must act better.

I cannot look into the face of one more child who has lost faith in himself and the world, I cannot look into the face of a child's father who is desperately trying to hold back tears; I cannot hear once again about how a school told a mother, 'we do not believe in dyslexia.'

As an iceberg is 90% underwater with only 10% visible; similarly, in dyslexia, we hear about the 10% like Max Brooks, who have made it. Let's not give up on the invisible 90% still underwater, asking, indeed, begging to be helped.

I am optimistic, once this committee is aware of the strong science of dyslexia, educators will want to align their practices and policies with 21st century science.

Recommendations

To bring education together with current scientific knowledge, the following are recommended:

1. First and foremost, schools must not be allowed to ignore, fail to recognize or deny the reality or diagnosis of dyslexia.

2. Schools, including teachers, principals and other administrators and parents should make every effort to use the word dyslexia since it has specific, highly relevant and explanatory meaning; science has provided its: definition; epidemiology; cognitive basis; neurobiological basis; developmental progression; long-term outcome. For dyslexia, knowledge of its cognitive basis indicates what symptoms to look for so that symptoms of dyslexia in the classroom (and at home) are noted and acknowledged rather than as currently happens, ignored or overlooked. This greater awareness and understanding of dyslexia and its impact will benefit both the teacher and student, both in the teaching of reading and in the climate and attitudes within the classroom.

3. Using the word dyslexia provides a common language facilitating communication among teachers, clinicians, scientists and parents.

4. For the student, the knowledge that he is dyslexic is empowering, providing the student with self-understanding and self-awareness of what he has and what he needs to do in order to succeed.

5. For students, knowledge that they are dyslexic also provides a community to join – they know they are not alone.

6. For the parent and teacher and importantly, the student, knowledge that he or she is dyslexic brings with it the information that the individual is not stupid or lazy.

7. Critically important is that schools must use evidence-based programs that have proven efficacy; research-based simply indicates that there are theoretical suggestions but does not provide evidence that the program is, indeed, effective. Evidence-based programs are akin to the level of evidence the FDA requires before a medication can be approved for use. Many, many theoretical, research based approaches, when tested in the field, prove to be ineffective. Our children's reading is too important to be left to theoretical, but unproven, practices and methods. We must replace anecdotal and common, but, non-

evidence-based practices, with those that are proven, that is, they are evidence-based.

8. Professional development programs targeted for teachers must provide <u>evidence</u> that the <u>students</u> of the teachers taking these programs actually improve in their reading performance. This is in contrast to some professional development programs which seem to improve teacher's understanding but not in a way that results in improvement in their student's reading performance.

9. Schools of education must ensure that aspiring teachers are taught evidence-based methods to teach reading and have monitored experience demonstrating that they are effective in implementing these methods.

10. Scientific evidence that reading growth is maximum in the very first few years of school and then plateaus together with new data indicating that the reading gap between typical and dyslexic readers is already present at first grade and persists means that students must receive evidence-based instruction at the start of their school experience and their progress carefully monitored. Waiting is harmful and not acceptable.

11. Given the rapid growth in reading in the very first years of school and the already present gap by first grade, it seems reasonable to encourage the creation of special charter schools for grades k-3 that focus solely on dyslexia. The goal is to reach children at-risk for dyslexia early on when reading intervention can be maximally effective and before the students fall further and further behind. At such specialized charter schools, such as the one, Louisiana Key Academy, attended by the children of a fellow panel member, the entire educational team from principal to classroom teacher to physical education instructor understand dyslexia, its impact on students in various situations and are on board to support the students throughout their day. Here, students learn and there is no bullying by students or frustration expressed by teachers who may not understand the impact of dyslexia. These schools can also serve as resources where teachers can come, spend time and learn about dyslexia, what it is and how it impacts a student and learn specific evidence-based methods for teaching reading to dyslexic students and how to best implement these methods.

There is so much more to tell; for those who have questions and want to know more, visit the Yale Center for Dyslexia &Creativity website: dyslexia.yale.edu or look at my book, "Overcoming Dyslexia," which discusses the scientific basis of dyslexia and how to translate this knowledge into practice.

Version 5

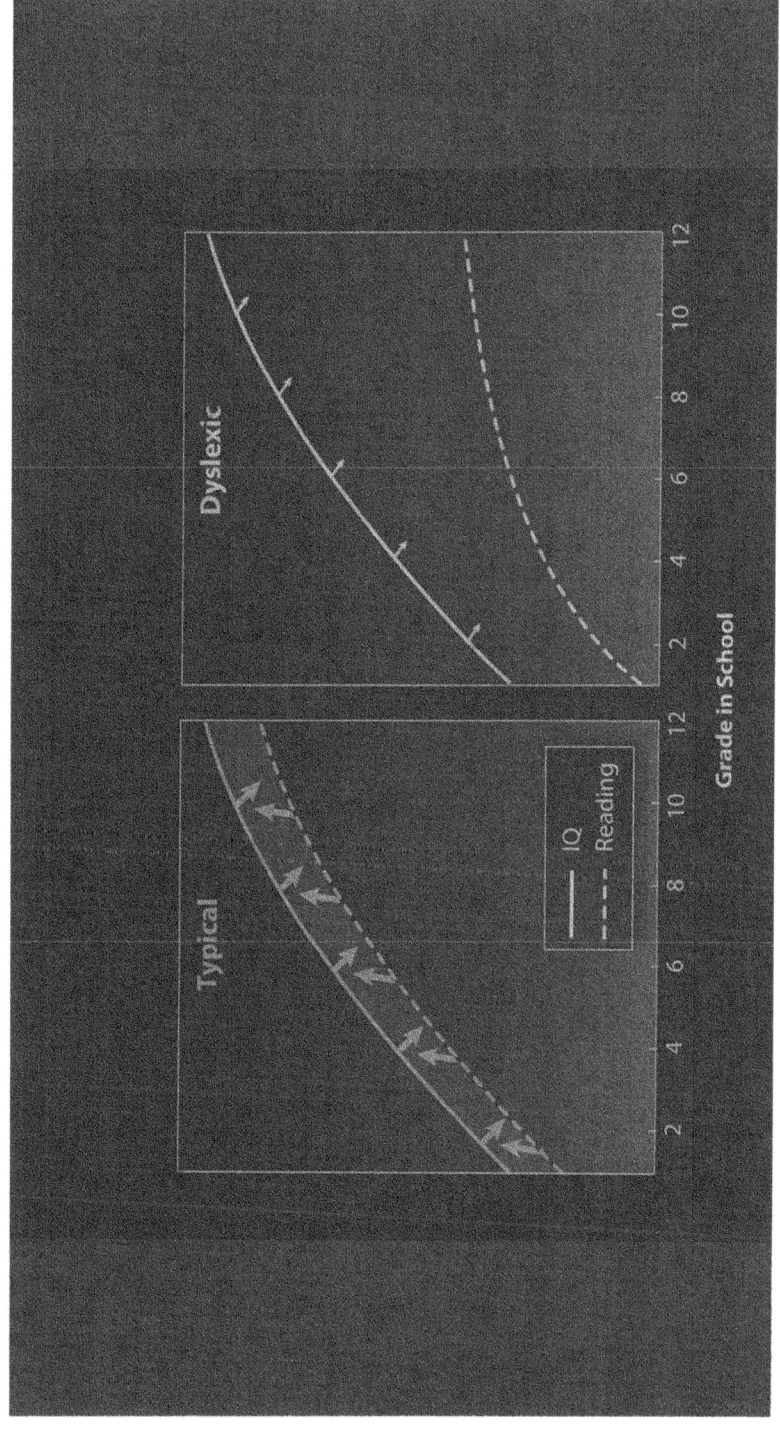

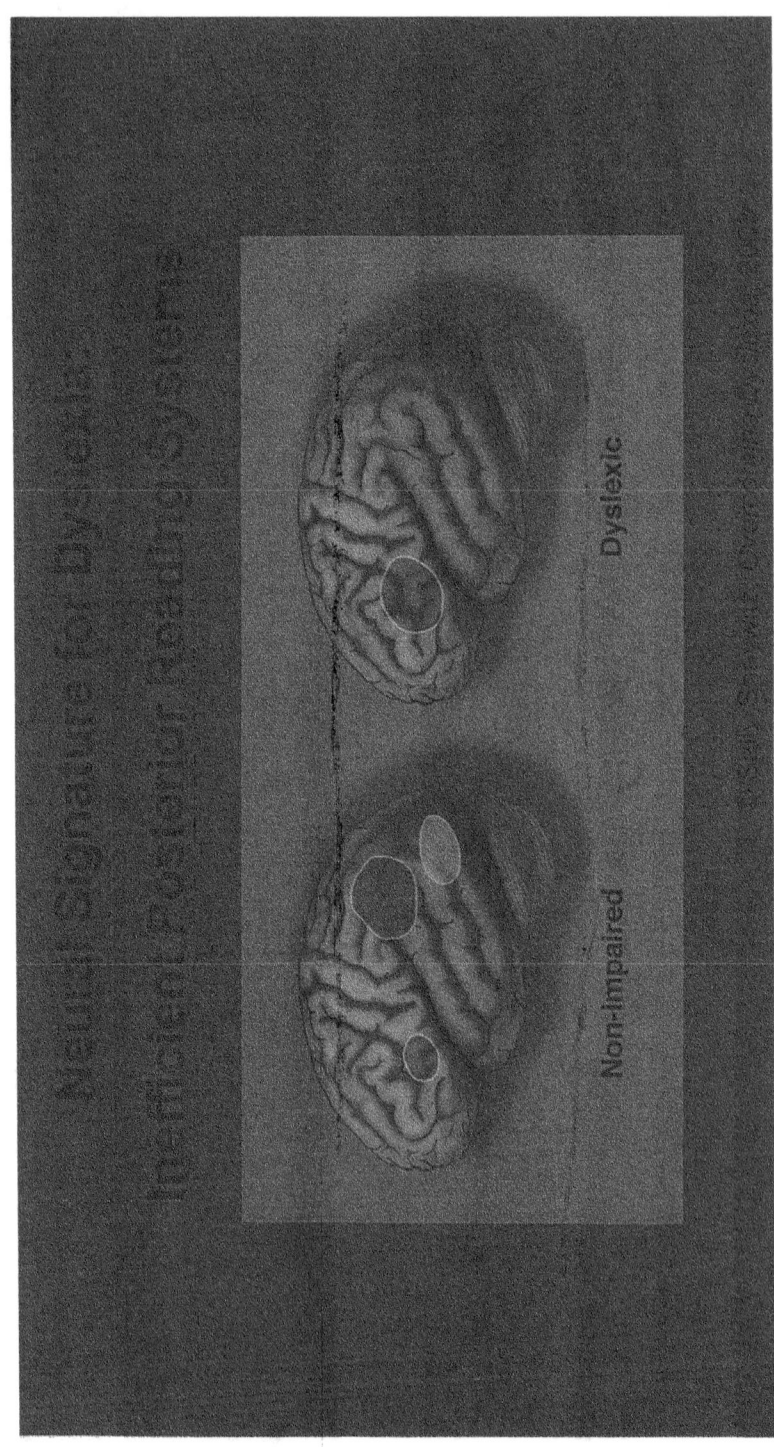

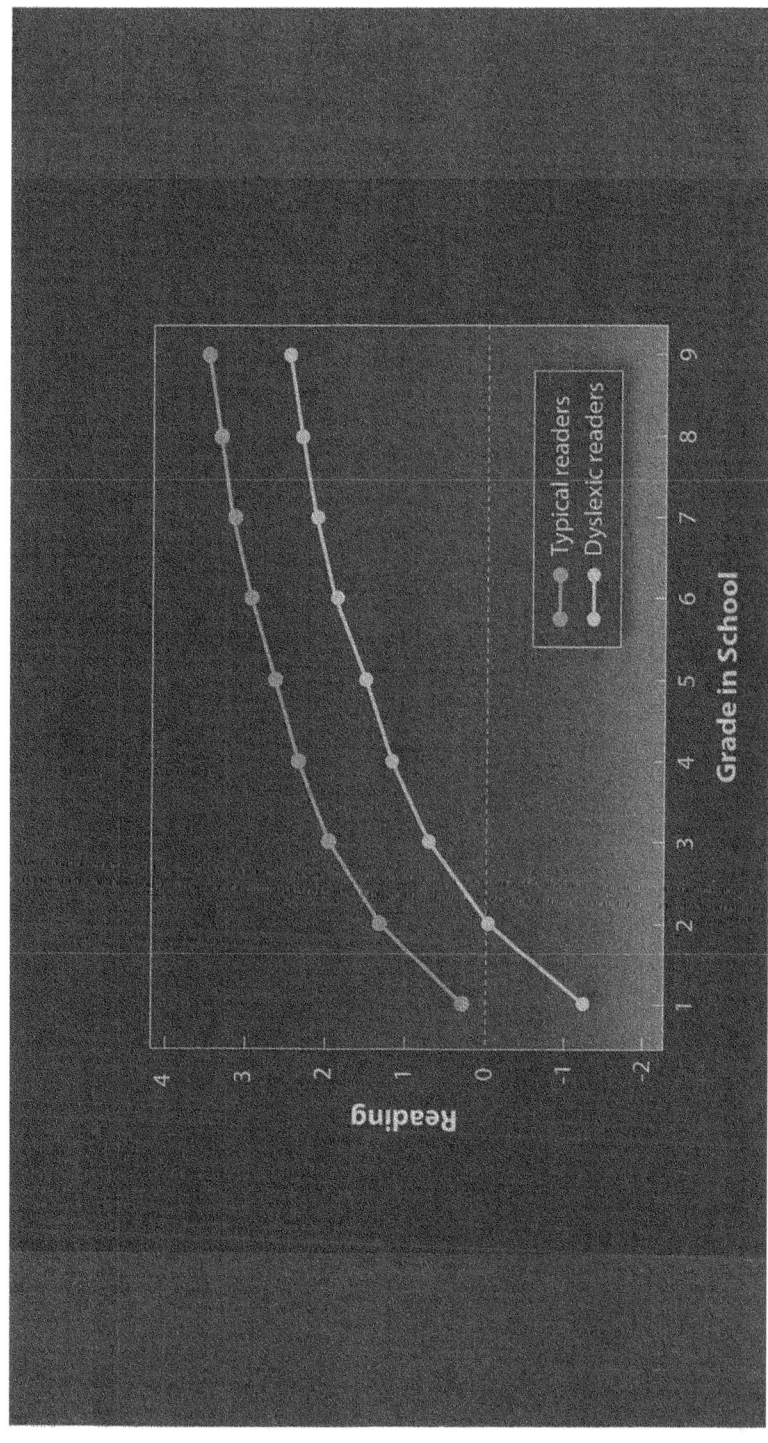

34

Sally E. Shaywitz, M.D. is the Audrey G. Ratner Professor in Learning Development at the Yale University School of Medicine and Co-Director of the Yale Center for Dyslexia & Creativity. Dr. Shaywitz' studies provide the basic framework and details for the 21st century scientific understanding of dyslexia. The author of over 250 scientific articles and chapters, her epidemiological studies provide current knowledge of the prevalence, gender composition, universality and persistence of dyslexia. Her functional imaging and cognitive research provides new insights aligning cognitive components of dyslexia with their underlying neurobiological substrate. Complementing her epidemiological, cognitive and neurobiological studies are ongoing longitudinal studies tracking a population-based cohort from kindergarten entry to mature adulthood. Her recent studies have provided the long-sought empiric evidence for the *unexpected* nature of dyslexia.

As a physician-scientist who investigates the scientific basis of dyslexia and also cares for people who are dyslexic, Dr. Shaywitz is dedicated to ensuring that scientific progress in dyslexia is translated into policy and practice. By invitation she has spoken at the World Economic Forum in Davos, Switzerland and here, in the US, presented the National Reading Panel report (mandated by Congress) to the US Congress.

An elected member of the National Academy of Medicine of the National Academy of Sciences, Dr. Shaywitz is annually selected as one of the *Best Doctors in America* and *America's Top Doctors*. Her awards include an honorary Doctor of Science degree from Williams College; the Townsend Harris Medal of the City College of New York; the Annie Glenn Award for Leadership from the Ohio State University; and the Distinguished Alumnus Award of the Albert Einstein College of Medicine. Dr. Shaywitz has served on the Congressionally-mandated National Reading Panel and the Committee to Prevent Reading Difficulties in Young Children of the National Research Council and, by Presidential appointment (President Bush, President Obama) on the National Board of the Institute for Education Sciences of the US Department of Education. She also co-chaired the National Research Council Committee on Gender Differences in the Careers of Science, Engineering and Mathematics Faculty. Dr. Shaywitz has also served on the Advisory Council of the National Institute of Neurological Diseases and Stroke (NINDS), the National Research Council Committee on Women in Science and Engineering and the Scientific Advisory Board of the March of Dimes.

Her book, the award-winning, critically acclaimed, "Overcoming Dyslexia" (Alfred Knopf, 2003) details fundamental scientific findings on dyslexia and how to translate this scientific knowledge into clinical practice.

September 15, 2014

Chairman SMITH. Thank you, Dr. Shaywitz, and Mr. Brooks.

TESTIMONY OF MR. MAX BROOKS,
AUTHOR AND SCREENWRITER

Mr. BROOKS. Thank you, Mr. Chairman.

I was asked here to talk about what it was like being dyslexic. I will say that for me, first of all, let me just say that I didn't prepare a statement because the last time I tried to read aloud from a prepared statement, my whole fifth-grade class laughed at me. Thank you.

So I am just going to be brief and speak from the heart. Dr. Shaywitz asked me to come here to talk about what it was like, and I will define it in one phrase that my teacher used to say to me in elementary school, which is ''You can do it. You just don't want to do it.''

Now, I grew up on the west side of L.A. in a very expensive private school. I had the best that the system could give me, and it wasn't enough, and I think the most important thing to discuss here is the psychological and emotional damage dyslexia causes. More than the learning disability is the blow to your self-esteem because once you are in that hole, it can take you the rest of your life to climb out. There is nothing more frustrating for a child to work twice as hard as the other kids but to do half as well. Eventually kids start to buy into the narrative, as I said, ''Well, maybe I am just dumb. I am clearly not lazy. I am not undisciplined.'' And when my teacher would say well, I am just going to whip you into shape, I would think well, yeah, that is exactly what I need is a whipping.

I was very lucky because I had one of the best moms ever, and I don't know how she knew about dyslexia in the late 1970s, early 1980s. She took that secret to her grave but somehow she knew about it. She made sure that I was diagnosed, tested and then she met with all my teachers and made them understand that me being the class clown and the troublemaker was my way of compensating for these horrible feelings of low self-worth. So she set in place therapies that helped me like taking an untimed test. An untimed test reduces the amount of anxiety that it gives a kid, because that is the problem with dyslexic kids. So much of it isn't the learning disability, it is the anxiety that it causes, which shuts down everything. So untimed test was important. Audiobooks—back in the day there were not many audiobooks so my mother took my whole school reading list every year to the Braille Institute for the Blind, had them read onto audiobooks so that way I listened to my school curriculum. Otherwise I wouldn't have passed.

Most importantly, she made sure that my teachers knew that I was trying and I was doing my best, and this was not some sort of voluntary screwing around. That helped me get through. And as a result, not only have I had success as an author, dyslexia has shown me to be a gift because I can't simply memorize facts and regurgitate them. I have to understand them. Because I have to understand them and understand the broader context in which they exist, it has made me a big-picture outside-the-box thinker, and that manner of thinker has gotten me invited to speak at places like the Naval War College and West Point, hurricane rehearsal of concept drills, the U.S. Army's Vibrant Response, and a few strategic studies groups that I don't think I am allowed to talk about here.

It is a gift, and we can turn so many of these kids around because that is the problem is that they start to believe well, if the system has no value with me, I have no value for the system, I am going to drop out, and they fill our streets and they fill our jails and they suck off the system for the rest of their lives because they don't feel they can contribute, and all we need to do is identify it and recognize it at the early level and we can turn these kids around, as we have all discussed. These are the creative thinkers. These are the engines of what we used to call Yankee ingenuity. How many Einsteins do we have sitting in our schools right now staring out the window because they think they have nothing to contribute? That is all it takes.

And look, I understand as a Member of Congress you have people coming to you every day saying listen, we have a problem and it will cost you $500 million a day. This can be solved, and it is one of the few problems you face as a Member of Congress that can be solved easily and relatively quickly.

Thank you.

[The prepared statement of Mr. Brooks follows:]

For me, dyslexia was nearly as bad as the feelings of anxiety, shame, and low self-esteem that it caused. For me, 'learned dependency' was the real enemy, the self-narrative that "I can't do this" can destroy children's learning potential for the rest of their lives. That was ALMOST me. I've spent the last 30 years unlearning the lesson that dyslexia taught me, that society has no use for me. How many other people were taught that lesson by ignorant teachers and bullying classmates? How many of them are in prison or on welfare, or simply divorcing themselves from the country around them? All we need is a little awareness and education, teaching the teachers to teach these kids that their curse is actually a blessing. My mother did it for me, she made sure I was tested, that I was tutored and she worked with my teachers to develop coping skills that I still use to this day. I learned by audiobooks and untimed tests. I got as much one on one instruction as I could. I've learned that the blessing of my dyslexia is just that because it forced me to be a critical thinker and truly understand what I was learning instead of just memorizing and regurgitating it for a standardized test. Dyslexia forced me to think outside the conventional cube and maybe that manner of thinking is why I've been invited to speak at the US Naval War College, at West Point, at various military and homeland security exercises and at a few strategic studies groups that I probably shouldn't be talking about here. And I'm just the tip of the iceberg. How many class clowns or troublemakers or dropouts could actually be the innovators of tomorrow if parents and teachers understood that dyslexia is both scientifically proven AND scientifically treatable? A little awareness and flexible teaching methods could unlock unlimited potential in these kids who now think they're losers. If we already have mandatory racial sensitivity training for our police, why not have mandatory dyslexia recognition training for our teachers? It's so simple, so easy, and when you look at all the other government programs designed to help citizens help themselves, it's probably the least expensive.

Max Brooks

BIOGRAPHY DETAILS

I was born in New York City in 1972; and I am the son of Hollywood legends Mel Brooks and Anne Bancroft and I was raised with my two feet planted firmly on the ground. I am recognized as successful author not only by the market place, but by critics and cultural observers. In my work, I am dedicated to challenging and encouraging systems and institutions to think outside the box to solve problems. Even when I write fiction, one of my major goals is to raise awareness on the issues of disaster preparedness, crisis management, and survival for the common reader—often under the thematic guise of a zombie apocalypse.

I have devoted much of my life to the study and development of systems and security, culminating in a genuine interest in the fundamentals and logistics that go into keeping our world safe from natural and man-made disaster threats.

Even in my loving and financially advantaged home, I still had some personal challenges to overcome. **As a child**, I was diagnosed with dyslexia. For me, dyslexia was nearly as bad as the feelings of anxiety, shame, and low self-esteem that it caused. For me, 'learned dependency' was the real enemy, the self-narrative that "I can't do this" can destroy children's learning potential for the rest of their lives. That was ALMOST me. I've spent the last 30 years unlearning the lesson that dyslexia taught me, that society has no use for me. How many other people were taught that lesson by ignorant teachers and bullying classmates? How many of them are in prison or on welfare, or simply divorcing themselves from the country around them? All we need is a little awareness and education, teaching the teachers to teach these kids that their curse is actually a blessing. My mother did it for me, she made sure I was tested, that I was tutored and she worked with my teachers to develop coping skills that I still use to this day. I got as much one on one instruction as I could; untimed tests were a life saver. I've experienced the downside of dyslexia by my slow reading and the bullying I've endured. I've also learned that the blessing of my dyslexia is just that because it forced me to be a critical thinker and truly understand what I was learning instead of just memorizing and regurgitating it for a standardized test.

Dyslexia forced me to think outside the conventional cube and maybe that manner of thinking is why I've been invited to speak at the US Naval War College, at West Point, at various military and homeland security exercises and at a few strategic studies groups that I probably shouldn't be talking about here. And I'm just the tip of the iceberg. How many class clowns or troublemakers or dropouts could actually be the innovators of tomorrow if parents and teachers understood that dyslexia is both scientifically proven AND scientifically treatable? A little awareness and flexible teaching methods could unlock unlimited potential in these kids who now think they're losers. If we already have mandatory racial sensitivity training for our police, why not have mandatory dyslexia recognition training for our teachers? It's so simple, so easy, and when you look at all the other government programs designed to help citizens help themselves, it's probably the least expensive.

After working for the BBC and then Saturday Night Live, I began writing *The Zombie Survival Guide* and then, what became a *New York Times* best-seller, *World War Z: An Oral History of the Zombie War*),which has been made into a major motion picture starring Brad Pitt.

Even though I am a best selling author and I had difficulty with reading in school, I never had difficulty with *thinking*, particularly with thinking *creatively*. While I am quite proud of having published four successful books—**The Zombie Survival Guide**, *World War Z*, *The Zombie Survival Guide: Recorded Attacks, and now THE HARLEM HELLFIGHTERS* —my ultimate goal was to challenge old ways of thinking and encourage mental agility and flexibility for problem solvers and leaders. I am told that my **unconventional thinking has even inspired the U.S. military to examine how they may respond to potential crises in the future.** '*Survival Guide'* **was read and discussed by the sitting Chairman of the Joint Chiefs** and, as I noted above, I have been invited to speak at a variety of military and crisis management conferences and symposia—from the Naval War College, to the FEMA hurricane drill at San Antonio, to the nuclear "Vibrant Response" wargame. By developing the dystopian mythos of a "zombie apocalypse" in film and literature, I have attempted to drive the dialogue as a thought leader on how to manage and coordinate emergency responses and to suggest better ways to prepare for crisis and conflict. Today, I see a crisis in the classroom and that is why I am here today.

In my work I am exploring the consequences of failed leadership, making the leap from mythos to American History with the release of my latest book, *The Harlem Hellfighters*. This book chronicles the little-known story of the first African-American regiment mustered to fight in WWI. They spent longer than any other American unit in combat and displayed remarkable valor on the battlefield. Despite extraordinary struggles and overt racism, the 'Hellfighters', as their enemies named them, became one of the most successful—**but least celebrated**—regiments of the war. **My goal, here too, is to drill down to help find what lessons can be learned from the triumph and tragedy of the Harlem Hellfighters. Success in the face of adversity, be it racism or DYSLEXIA LIKE I HAVE.**

Using fictional metaphor and historical events, my goal is to prompt serious discourse on large-scale problem-solving, and to explore new ways to attack old problems and new concerns. For example, whether I am on stage speaking or writing, my goal is to tackle the tougher questions, like what are the threads that hold society together, and what's really at stake when those threads are stressed, loosened, or torn. From the basic responsibilities of civilians and corporations to the over-arching role of government and the military, I try to shed light on what each can—and should—be prepared to do, to not only survive, but thrive in the face of real-world threats, disaster, and the unexpected.

Chairman SMITH. Thank you, Mr. Brooks. Ms. Antie.

TESTIMONY OF MS. STACY ANTIE, PARENT AND ADVOCATE

Ms. ANTIE. My story is simple. I am not a medical doctor or a Ph.D. I am just a mom that is here to try to explain the daily challenges that my 9-year-old child faces, who struggles with dyslexia, with the hope of bringing awareness for every child that has it, whether they are diagnosed or undiagnosed.

Dyslexia often prohibits my son from having the ability to match sounds with the alphabet. This impacts his reading, his spelling and his speaking. It is not a sign of poor intelligence or laziness. In fact, my son is exceptionally bright and very intuitive. He is creative. He is very tenderhearted and loving.

In kindergarten, I noticed that something was a little bit different with my son. He had a harder time with things than other children. He was not able to rhyme or even understand the concept of rhyming. This continued throughout kindergarten. I brought him to a reading clinic over the summer between kindergarten and first grade. All the teachers told me the same thing: he is a boy, he will get it, boys tend to get it a slower but he will have it by Christmas. Christmas came, Christmas went; he still didn't understand the concepts.

He started first grade, and first grade is where our real troubles started. He became very anxious about going to school. He didn't want to read out loud in front of all the other children. He became very frustrated. The 20 minutes of homework that he would have had would turn into two to three hours every night. He would have to read where it would normally take a few seconds, would take him five to six minutes because he had to sound out every letter of every word and then try to blend it together, which he couldn't hear the sounds to blend them. From there, it turned into tears for both he and I because we sat there for hours working on his homework until he went to bed.

In mid-October of his first-grade year, his teacher said, ''I am seeing some red flags, you might want to have him tested for dyslexia,'' which I did. I struggled with my insurance company to have him tested, and they told me that he couldn't be tested until June, which was eight months away. The longer that I waited for him to be tested, the further behind he was going to get, so I decided to pay out of pocket and have him tested, and he was given the diagnosis in December of 2011 that he was dyslexic and he had developmental coordination disorder.

He started speech therapy and occupational therapy a few times a week. He also got tutoring. He had already failed first grade, because if you fail reading for the first grade, you fail the entire year, so we knew he had failed by the third semester of first grade. He would come home every day and ask me, Mom, why am I so stupid and no one else is? And as a parent, how do you look into your 7-year-old child's eyes and reassure him that he isn't stupid, he just thinks a little bit differently than other people. He would never read out loud, and his self-esteem plummeted. He became very introverted and he never wanted to do anything outside of the house.

The next school year in 2012, he started first grade again. He had a B average, because he had already had all the material previously. He still struggled a little bit in reading, and he went to a reading interventionist every single day for 30 minutes a day in addition to all the therapy we were having. At the end of the school, there wasn't a huge improvement in his reading abilities, and I thought, why am I paying all these thousands of dollars in tuition when the school really can't help my child. It doesn't have the resources to help him. They tried to compensate for his needs but they weren't able to correct anything. He didn't fit into the "normal" mold that all the other children did, and at that time they— I am sorry—they didn't want to change their mold for one child.

I found out about a new charter school at the end of that year, Louisiana Key Academy. I applied for him, and he was immediately accepted since he already had the diagnosis of dyslexia, and for the first time he was excited about going to school because he was along with other children who felt the exact same way he did. He was able to understand that they have trouble reading just like we did. His self-esteem picked up a little bit each day. He felt like he belonged. He went from a classroom size of 32 to a classroom size of 16. The classroom size broke to six at reading, which is instrumental in him learning because he was able to get individual attention along with learning from a small group.

Louisiana Key uses a systematic evidence based curriculum called Neuhaus, and this teaches them how letters work to form words, and he started to hold his little head a little bit higher and stopped referring to himself as stupid.

The pivotal point in this journey with me happened in January of this year. I was helping my son, his brother, who is 7, and Coleman, who is dyslexic, came up to me and he said, "Thomas, you know what to do. It is a CVC word. That means you slice the E and you make the A a macron," and I was looking like he was speaking Chinese to me because I understood nothing about it, and he said, "Mom, let me break that down for you. You just make it long," and I said okay, so I pointed to another word and he immediately showed me how to do that one, and I told him how proud I was of him and he said, "I understand how to decode words now. It just makes sense."

At the end of last school year they had to do a project where they had to read a presentation. They had to become a famous American. He chose Steve Jobs because my son is great at videogames and Steve Jobs makes the absolute best thing in the world for my son to play videogames on, but as we learned, he was dyslexic also. We used that as a tool to help all the children learn that if they work through this and they work together and they work hard and they stay determined, they can do anything. And as he went for his presentation, I held my breath, and he stood before his class and he read in front of me and all the parents and all the children, and he has never done that ever.

So I thank you for this opportunity just to explain all the challenges that these children really do go through, and as a family, everybody is affected, not just the child.

[The prepared statement of Ms. Antie follows:]

My story is simple. I am not a medical doctor or a PhD. I am a Mom who is here to help explain the daily challenges of my nine (9) year old child who struggles with dyslexia with the hope to bring awareness for every child with it (whether diagnosed or undiagnosed).

Dyslexia often prohibits my son from having the ability to match sounds with the alphabet. This impacts his reading, spelling and speaking. It is not a sign of poor intelligence or laziness. In fact, my son is exceptionally bright and intuitive, creative and very tender hearted and loving.

My son attended daycare and then went to Pre-K at four (4) years old. At five (5), he started Kindergarten at a private school. About halfway through his Kindergarten year, I noticed that he was unable to do some things that other children did. Most things were harder for him than other children. He would write several letters backwards, which the teachers said was normal at this age. In addition, he could not understand the concept of rhyming. We would practice every day, played games with it, and even rhymed in the car. Yet, he still was unable to do it. I would say the word "cat". Where other children would rush to say bat, mat, or sat, my son would say "tab". He would only hear the last syllable and start the next word with that. He read at a very slow pace and would have to sound out every letter in the word, often being unable to hear them when he said them together. If I sounded them out loud, he would occasionally be able to blend them and make the correct word. I was concerned at the end of the school year and met with his teacher. She assured me that there was no need to worry. That he was a boy and boys tend to take a little longer to read but he would get it on his own time.

I decided, on my own, to place him in a reading program through Louisiana State University over the summer between his Kindergarten year and First grade. It was a six (6) week program geared to make stronger readers. He was assessed after two weeks in the class and placed in the lower reading group. We continued to work and he did not improve much, if any. I met with this teacher and she assured me that he was trying and would get it on his own time.

He started first grade. The first two weeks were great, then, the real trouble started. The frustration and anxiety increased tremendously with him. His twenty (20) minutes of homework would easily turn into two (2) hours. He brought home a worksheet with sentences to read every night. One sentence would take us up to five (5) or six (6) minutes instead of a few seconds. It was very hard to find the patience to listen to him sound out letters and then try and blend them into a word. If he would have trouble with a word in a sentence that wasn't easily sounded out, I would help him with it. When the next sentence contained the same word, he acted as if he had never seen the word before in his life. I would get frustrated and explain to him that he just read it no more than ten (10) words earlier. I could not understand why he was unable to see that was the same word and just say it. This led to crying (on both of our parts), each and every night, from the time we started homework until the time that he went to bed. At an appointment with his teacher in early September, she assured me that he was a boy and boys took longer to read sometimes but he should have it by Christmas.

In the middle of October, his teacher asked to meet with me. She stated that she was seeing some red flags and asked if I was open to having him tested. She was very honest that she did not have a degree to assess him but he was showing some signs that had caused her some concern. I went through my insurance company and made an appointment with a local doctor for the end of October. The doctor was not at the office for my appointment, yet spoke to me through Skype. His assistant ran some tests on my child. In mid-November, I returned for the results. I still was unable to meet the doctor, yet, my son was given a generic diagnosis of dysgraphia from the handwriting samples that I provided from school.

I contacted my insurance company and they said that it would be June before I could get in to see anyone else. Frustrated, I didn't know what to do next. Every minute I spent not getting help for my son, was a minute that he was falling further and further behind. Completely distraught, I decided to have him tested at a psychological facility and pay for the testing out of pocket. He was tested at the end of December 2011, by a PhD, who actually met with my son and me in person. In the middle of January 2012, my son was given the diagnosis of Dyslexia and Developmental Coordination Disorder. It was recommended that he start speech and occupational therapy. We also started with a tutor three times a week for help in reading.

I received the report at the end of February and met with the school at the beginning of March 2012. By this date, my son had already failed for the year because if a first grader fails reading, he fails the year. We made his IEP for the next school year. It consisted of testing in another room, having everything read to him except his reading test, and he started seeing the Reading Interventionist every day at school.

My son would come home every day and cry and ask, "Mom, why am I stupid and no one else is"? How, as a parent, do you look into your seven (7) year olds eyes, and reassure him that he isn't stupid but that his brain just acts a little different from everyone else's? His self-esteem plummeted. He would never read a book out loud to anyone but me. He became very introverted and never wanted to do anything outside of the house.

The next school year, August 2012, he started first grade again. His grades were much stronger this year. He maintained an A/B average with an occasional C; however, reading was still in the low B's. He was having occupational therapy a few times a week after school. He had speech therapy once a week after school. He was meeting the Reading Interventionist every day at school for thirty (30) minutes and he was still being tutored three times a week after school. I met with the interventionist towards the end of the school year and she told me that she saw only the slightest improvement with his reading, although she would have expected it to be a little more since she worked with him every day.

At this point, I was dumbfounded that I was paying thousands of dollars in tuition and this school really didn't have the resources to help my child. It wasn't that they didn't try to compensate for his needs; however, my child didn't fit into the "normal" mold of students in his class. The school was not going to change their teaching habits for one child. So, they tried to mold my child to learn like every other student although it was impossible for him.

On May 23, 2013, I found out about a new school that was being started in Baton Rouge, Louisiana. It was a charter school for dyslexic students. I immediately contacted them to see how I could apply. To qualify, the child had to have a type of dyslexia or reading issue with an evaluation to support it. Since my son was previously diagnosed, I applied and my son was accepted. I honestly had some reservation because I was very unsure what to expect. In addition, he had to make new friends just the year before and he was already so introverted and now I was making him make new friends all over again.

In August 2013, my son started at Louisiana Key Academy. A few days into the school year, he came home and was excited to talk to me because he found out that a boy in his class had trouble reading, too, and so did a girl in his class. For the first time in his life, he didn't feel like something was wrong with him because everyone in this school was just like him. His self-esteem picked up a little each day.

In addition to feeling like he belonged for the first time, the classroom size was dramatically different than we had previously experienced. He went from a classroom of 32 (thirty-two) children, at the prior school, to a classroom size of 16 (sixteen). During reading, his class size dropped to 6 (six) children. He was able to get individual attention while also learning from the other children in his group.

Louisiana Key Academy uses a systematic evidence based curriculum called Neuhaus. The teachers and reading interventionists are trained to teach this evidence base curriculum. The more knowledge he gained through the Neuhaus program about how letters work to from words, the higher he held his head. He stopped referring to himself as "stupid"!

The pivotal point in this journey with my son happened around January of this year. I was helping my other son, who is seven (7) with his homework while my son, now nine (9), with dyslexia was sitting at the table working on homework, too. My seven (7) year old was having some trouble with a word and my nine (9) year old jumped up and said, "If it is a vcv, you slice the E and put a macron over the first v". I looked a little confused and he looked at me and said, "Macron means that you make it long, Mom". I immediately pointed to another word and he explained that one to me too. I told him how proud I was of him and he said, "It is easy now that I understand how to decode words. "It makes sense now Mom."

At the end of last school year, my son had a project where he had to be a famous American. Each child dressed as the person and had to present an oral report. My son chose Steve Jobs. While my son chose Mr. Jobs because he loves video games and Mr. Jobs created my son's favorite way to play them, we learned that Mr. Jobs was dyslexic. We were able to use his report as a way to show each child in his class that they can succeed – it just takes hard work and dedication. As I held my breath, my son stood before his entire class and their parents and read his report out loud (Something that he has never done – EVER).

With all the programs and options that I have tried, I finally see something that is working for my child and I cannot express the gratitude that I feel. Although my son has overcome so much, he still has daily hurdles that he must conquer. My son still has anxiety about school and some social settings and probably always will. He cries occasionally when he wakes up and knows that it is a school day. He continues to be a poor speller and, at times, he still has difficulty getting words out. However, he finds the strength to push through each challenge and succeed.

Reading difficulties are the most common cause of academic failure or underachievement in our society. An inability to read affects every aspect of a person's life. Luckily, for my son, I have been vigilant in trying to find a program that will help my child succeed, as I am sure every parent here today would do.

In my opinion, early identification is the key. The longer we wait for a child to be diagnosed, the more valuable time is wasted that could be helping children before they fall further and further behind. Smaller groups and class sizes are invaluable, especially for the children who do not fit the standard learning "mold". I also feel that children need to around other children who struggle the same way that they do they don't feel ostracized and have lower self-esteem. One thing that each of us need to remember is that our children are the future. We might not have a cure for dyslexia, but together, we can find a solution.

BIOGRAPHY

STACY ANTIE serves as the Administrative Assistant to a Director of a State Agency. She has become vigilant in finding her nine (9) year old son, who was diagnosed with dyslexia, the best type of treatment to ensure his success. She has been spreading her son's story for the past year to help other families in her area who feel helpless.

Chairman SMITH. Thank you, Ms. Antie. Dr. Eden.

TESTIMONY OF DR. PETER EDEN,
PRESIDENT, LANDMARK COLLEGE

Dr. PETER EDEN. Thank you, Mr. Chairman.

Students with dyslexia are certainly disadvantaged. I want to focus on students at the college-university level.

Hundreds of thousands of students with dyslexia are in colleges and universities right now, and there are many, many more with other learning disabilities and learning difficulties such as ADD or autism spectrum disorder. Most students with dyslexia cluster in two-year colleges, and while the majority received accommodations in high school, very, very few receive accommodations in college. When they do, it is typically extended time for a test.

One rarely finds comprehensive support models and programs in higher education, let alone a dedicated program like Landmark College, which focuses only on students with LD such as dyslexia. This is only going to get more challenging when one considers the reality of online teaching and learning, which is here to stay. The standard modalities we use in education today, in online education, will only exacerbate these challenges for students with LD such as dyslexia. We need more adaptive learning elements. We need more quality assurance when teaching and learning in an online milieu.

In terms of some innovative educational practices and efforts, the principle of universal design, UD, or UDL, gives us great promise and hope, and UD is a principle where one engineers the learning environment to anticipate the neurodiversity in the classroom, to anticipate the heterogeneity of learners, of students and learning profiles. UD ensures that there are multiple means of presentation, student responses and engagement, and UD can be applied in conventional or in online settings.

Another innovative educational practice involves mobile technologies. Already today, I have heard people mention that students are using iPads and they certainly have smartphones. Greater than 90 percent of college students own a smartphone. A great number own an iPad, and when do own an iPad, greater than 90 percent of these college students use the iPad for learning.

We need to meet these college students where they exist, which is online and using a mobile device and develop assistive integrated technologies so we have a more ubiquitous ecosystem for them, to specialized rooms. We can remove the stigma. We need to focus on elements which use native software in our smartphones and iPads for teaching and learning and it will always be available to our students.

Also, cognitive training is another innovative educational practice where we explore, for example, patterns of learning through markers, cognitive, physiological and other markers, via videogame activities. This is one way to leverage the fact that thousands, millions of learners play videogames. We can assess big data, get our hands on exhaust data, learn how they learn within a videogame and seek adaptive and customizable learning activities.

In terms of Landmark College, we offer 2- and four-year degree programs including those in STEM. We serve only students with some learning disability or learning difficulty. This is a dedicated model. We have 500 students. Every student has some learning challenge. We use UD principles and integrated technologies. We provide careful placement and curriculum tracks for students with dyslexia. We have a hidden curriculum of support with our resources. Our retention rates are high. The ultimate B.A./B.S. graduation rate for Landmark College students is 70 percent, which is higher than the national average for all students, let alone students with LD. We juxtapose research and innovation with teaching and learning.

Now, at Landmark College we have the Landmark College Institute for Research and Training, LCIRT. LCIRT has received recent funding through LDFA for iPad app development between students and faculty, and also two recent NSF awards. I will try to summarize those now. One, NSF REAL, Research in Education and Learning. This is where we will investigate the efficacy of instructor presence in synchronous elements for online learning in STEM content for students with LD including dyslexia whether or not instructor presence and response immediacy pays off with student outcomes. Also, another grant recently received by the college is NSF data-intensive research to improve STEM teaching and learning, collaboration with MIT and TERC, and this is Revealing the Invisible grant funding. We are using the game vehicle to study engagement, eye tracking of students playing videogames, attention, memory, and implicit understanding of Newtonian physics in this effort. Again, we could capture huge amounts of data given the number of students involved in gaming in understanding how they learn in that environment. It also provides for educational data mining and an understanding of machine learning.

In summary, LD such as dyslexia provides a true barrier to learning, education and employment. There is a huge untapped population of potential workers, for example, in the STEM field. Innovative educational practices that are scalable that we can disseminate are needed, particularly with the online education realities. We use technology to discover better teaching and learning platforms and to understand neurodiverse students and to provide ubiquitous tools for success in college and in careers. Research and support is increasingly focused in this area. We are grateful, but much, much more is needed, and remember: advances in this area for students with dyslexia and LD, they will provide improvements that are good for all learners, for every learner, and that is why this is so important.

Thank you.

[The prepared statement of Dr. Eden follows:]

OFFICE OF THE PRESIDENT

House of Representatives Presentation Testimony

Peter Eden, Ph.D. President

Summary. Landmark College opened in 1985 to students with language-based learning challenges such as dyslexia. Unlike most colleges and universities, we did not – and still do not - assume that our students learn the same way, or the way neuro-typical individuals learn. We continue to serve only students with a learning disability/difficulty ("LD" writ large for this report) such as dyslexia, but also students with attention deficit disorder (ADD) and/or autism spectrum disorder (ASD). Until two years ago Landmark College offered 2 year associate degree programs but now we offer 2- and 4- year programs in STEM and other disciplines (as well as a graduate-level certificate program for educators and professionals). We are unique among other institutions of higher education in this regard i.e. we do not offer just an optional resource or program to students with LD but rather we *only* serve students with LD. This year we have the highest enrollment (about 500 students) in our history and we are constructing a new Science and Innovation Center on campus to support our students and our grant-funded research.

We understand that dyslexia is a language-based processing disorder that can affect an individual's ability to read, write, spell and speak and often this impedes social interactions and self-esteem. Perceptual accuracy, phonological impairments, slower auditory processing speed, and short term memory are all areas of deficit for a dyslexic, and thus impede various school-related abilities that rely on both receptive and expressive processes. Dyslexia is the most prevalent specific learning disability, comprising at least 50% of the LD population. As described by the International Dyslexia Association, "as many as 15–20% of the population as a whole have some of the symptoms of dyslexia, including slow or inaccurate reading, poor spelling, poor writing, or mixing up similar words. Not all will qualify for specific resources and assistance in education, but they are likely to struggle with many aspects of academic learning and are likely to benefit from systematic, explicit, instruction in reading, writing, and language." In terms of incidence in higher education a conservative estimate is that 1.2% of college population in the U.S. suffers from disorder. According to the National Center for Education Statistics, there are 21 million students enrolled in U.S. colleges and universities; therefore, there may be at least 250,000 college students with dyslexia. The majority (about 83%) of these learners are not obtaining post-secondary accommodation for their LD. Retention and graduation rates for students with LD such as dyslexia are below national averages for all learners, and employment rates suffer.

Landmark College blends universal design for learning (UD) principles and integrated technologies within a comprehensive model of support in and out of the formal learning environments, starting with initial assessment and curriculum track placements. Our student outcomes are strong, e.g. our students ultimately earn baccalaureate degrees at a rate higher than the national average for all learners. We conduct discovery and applied research in the field of LD (including recent support from NSF to LCIRT, the Landmark College Institute for Research and Training) and we are quickly developing new platform in online/web-based programming to better serve students with LD and to reach educators struggling to keep up with the heterogeneity of learners in their classrooms.

I.   **The Field of Dyslexia and LD**
      The field of dyslexia is being driven by a confluence of three strands of thinking: (1) innovative educational practices, (2) new medical research about dyslexia and the dyslexic brain, and (3) amendments and reauthorization of legal mandates re. equal access and non-discriminatory policies.

      Innovative Educational Practices. One approach is the paradigm of universal design (UD; also Universal Design for Learning or UDL). Some helpful links: www.cast.org and www.udi.uconn.edu. UD is significant as an alternative to the current Accommodation model which can stigmatize students with disabilities as well as creating barriers to obtaining services. UD obviates the need to identify students with dyslexia by building supports, such as accessible text and universally available lecture notes, into the curriculum.

      Mobile technologies offer advantages as well. For example, modifying content according to user preference and need in a ubiquitous environment, without the need to access specialized offices or study labs (e.g. digitized text available on a smartphone, with built in text-to-speech and modifiable font size and format).

      Another area of innovative educational practice is the area of *cognitive training*, particularly in the arena of video games and gamification. Landmark College is working with Akili (an associate of the UCSF Gazzaley lab) on the benefits of video games on students' attention. LCIRT at Landmark College, together with MIT and TERC, Inc. has recently received NSF funding for a 2.5 year project that seeks to explore patterns of learning examined through psychological, physiological and cognitive markers, while students engage in science-based video games.

      New medical research. We now know that dyslexic symptomologies can be attributed to more than a deficit in phonological processing (i.e. speech to print association). The brain is a dynamic ecosystem [1] and new understanding of the neuroplasticity of the human brain is creating a more hopeful landscape for dyslexia than ever before. We now know that the pathway to dyslexia is multi-factorial and different subgroups within this umbrella term need different solutions for success.

      In their book titled the *Dyslexic Advantage*, Drs. Brock and Fernette Eide (2011) point out that the dyslexic brain is better viewed as a trade-off model rather than a deficit model. They talk of the MIND strengths where, for example, a brain that is especially attuned to viewing information in 3-D images (especially advantageous for architects), may also engage in reversing letter and numbers (symptoms often associated with pure dyslexics).

      Legal mandates. The *Individuals with Disabilities Education Act (IDEA)* governs how states and public agencies provide early intervention, special education, and related services to children with disabilities. It addresses educational needs of children with disabilities from age 3 to age 18 or 21 (whichever comes earlier in graduating from school). The focus of IDEA is academic success through programs that support school students in accessing the general curriculum. IDEA ensures free appropriate public education (FAPE) to all and applies to all institutions receiving federal funding.

      Americans with Disabilities Act Amendments Act of 2008 (ADA AA). The ADA is a non-discrimination law not a special education law like IDEA. The ADA will be 25 years old in 2015. The ADA enables society to benefit from the skills, talents and purchasing power of individuals with disabilities and facilitates fuller, more productive lives for all Americans. It defines an individual with a disability as someone who has a physical or mental impairment that substantially limits one or more major life activities or bodily functions, has a history or record of such impairment, or is perceived by others as having such impairment. The ADA does not specifically name all of the impairments that are covered.

Any institution receiving federal funding falls under the purview of the ADA. And recent thinking after the 2008 Amendments includes that the burden of proof should not rest on the individual alone in establishing eligibility as an individual with a disability and need for reasonable accommodations, and that a request for proof of disability cannot be burdensome, but institutions can still require proof.

Higher Education Reauthorization Act. The Higher Education Opportunity Act (Public Law 110-315) (HEOA) was enacted on August 14, 2008, and reauthorizes the Higher Education Act of 1965 as amended (HEA). The 2008 reauthorization of HEOA included provisions to increase postsecondary education opportunities for students with disabilities by creating new and sustaining existing programs to increase access, recruitment, retention and completion rates; identify and promote effective transition practices; increase access to instructional materials, and disseminate best practices related to postsecondary students with disabilities.

One of the best practices articulated in the HEOA is UD (or UDL). The Higher Education Opportunity Act of 2008 includes in its language both a formal definition of UD and guidelines for providing UDL training to future teacher educators. The HEOA defined UD as "Universal Design for Learning means a scientifically valid framework for guiding educational practice that (A) provides flexibility in the ways information is presented, in the ways students respond or demonstrate knowledge and skills, and in the ways students are engaged; and (B) reduces barriers in instruction, provides appropriate accommodations, supports, and challenges, and maintains high achievement expectations for all students, including students with disabilities and students who are limited English proficient [HEOA, P.L. 110-315, §103(a) (24)]."

In other words, under the HEOA, universal design for learning has been explicitly defined and is integrated into the programs that are part of the law. The law states that recipients of 'teacher quality partnership grants' and 'teach to reach grants' must offer preparation programs that enable teachers to understand and use "strategies consistent with the principles of universal design for learning" [P.L. 110-315, §202(d)(1)(A)(ii)], and "to integrate technology effectively into curricula and instruction, including technology consistent with the principles of universal design for learning" [P.L. 110-315, §204(a)(G)(i)].

The inclusion of UD in the reauthorization of the HEOA demonstrates its escalating importance in the education field. UDL concepts and practices are not yet broadly integrated into all education policy and this is an important need. *Landmark College has been implementing UD/UDL principles in its pedagogical practices for a long time and well before UDL was adopted by the HEOA.*

II. **Realities and Challenges Relating to LD and Dyslexia in Post-Secondary Education**
The 2005 U.S. Survey of Income and Program Participation (SIPP) suggests that 1.8 percent of the U.S. population aged six years or older has been diagnosed with a learning disability, a figure which translates to approximately 4.67 million Americans.

- o 2.4 percent for those aged 6-11 years
- o 3.4 percent for those aged 12-17 years
- o 2.7 percent for those aged 18-24 years (the typical age of college students)

About one third of students with disabilities have a diagnosis of LD, ADHD, and/or ASD (United States Government Accountability Office, 2009)[2] with numbers in all three categories on the rise.

Dyslexia is the most prevalent specific learning disability, comprising at least 50% of the LD population. This suggests that 1.2% of college population in U.S. suffers from disorder. According to the National Center for Education Statistics, there are 21 million students enrolled in U.S. colleges and universities. Therefore, *there are approximately 250,000 college students with dyslexia.*

Impact on Higher Education Outcomes. Findings gathered by the National Center for Learning Disabilities indicate that students with LD are less likely to gain access to higher education. In 2011, just 68 percent of students with specific learning disabilities (SLD) graduated with a regular high school diploma. Close to half of secondary students with LD perform more than three grade levels below their enrolled grade in essential academic skills (45% in reading, 44% in math).[3]

Students with LD cluster in two-year colleges; young adults with LD attend two-year or community college at more than double the rate of the general population. About 10% of students with LD are enrolled in a four-year college within two years of leaving school, compared with 28% of the general population.[4] Indeed, young adults with LD attend four-year colleges at half the rate of the general population.[5] In terms of graduation rates, *students with LD who do enroll in 4-year colleges are less likely to graduate within 6 years.*[6]

Overall, people with disabilities (Note: all disabilities) complete college at a statistically significant lower rate than people without disabilities. Those who do complete college have a persistently lower rate of employment irrespective of the level of degree attainment (associate, bachelor's, and higher) (Bureau of Labor Statistics, 2012).[7]

Graduation Rates of Students with or without LD

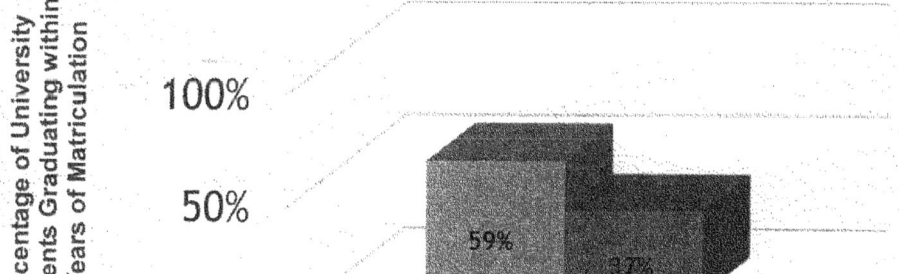

Other studies have found 6-year graduation rates for students with specific learning disabilities (over half of whom have dyslexia) as low as 29%[5], and one study has reported a 34% 8-year graduation rate for students with LD.[8]

Employment: Partially as a result of lower levels of success in higher education, students with LD are less likely to be gainfully employed.[9]

56

Employment and Labor Force Participation of Adults with or without LD

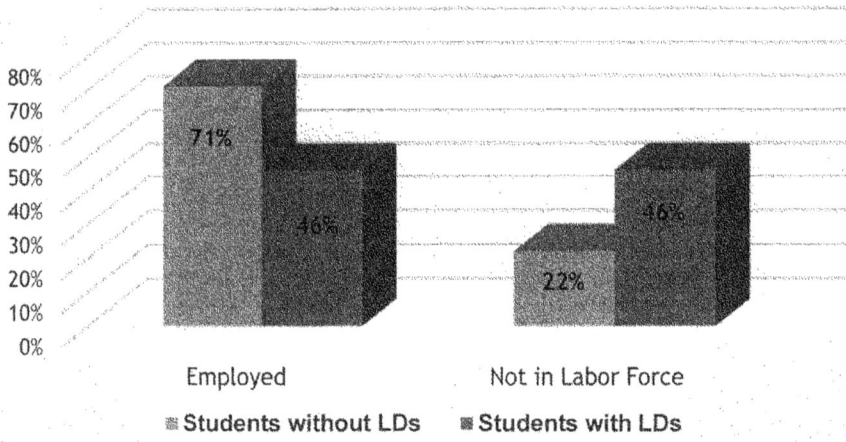

- ▨ **Students without LDs**   ▨ **Students with LDs**

Availability of Higher Education Support Services for Students with Dyslexia and other Specific Learning Disabilities. Only 17 percent of young adults with LD received accommodations and supports in postsecondary education because of their disability, compared to 94 percent in high school[8]. This indicates that 83% of students with dyslexia, or approximately 200,000 students, are *not* obtaining post-secondary accommodation for their LD. By contrast, 35% of students with ASD or traumatic brain injuries receive accommodations and supports.[10] Approximately 92% of students with LD receiving assistance with school work found this assistance to be useful.[9]

The rate of receiving accommodations and supports in postsecondary schools for those who had disclosed a disability ranged from 58 percent at 4-year colleges or universities to 64 percent at vocational, business, or technical schools and 76 percent at 2-year or community colleges.[11] Nearly 80% of support provided consists of additional time for tests. Only 37% of students with LD receive adaptive technology or tutoring.[12]

Only 69 U.S. colleges and universities provide comprehensive support programs for students with LD (Landmark College was the first, and currently one of only two, with dedicated overall program). Most located in 4-year colleges.[13] Such programs often combine academic support with elements such as:
- o   Weekly workshops on topics such as time management or peer interaction
- o   Development of individualized learning plans
- o   Weekly meetings with counselors

The most common tuition model appears to be the assessment of a standard fee per semester or per year, typically in the range of $1,500 to $3,500 per semester.[12] For Landmark College, all support systems and elements are included with the overall tuition and fees (not extra charge).

Are individuals diagnosed with dyslexia given opportunities to succeed? Opportunities available to students with dyslexia today are much more inclusive and expansive than just a decade or even 5 years ago, but more can and needs to be done. Both secondary and postsecondary institutions should consider an across-the-life-span approach to supporting students with LD such as dyslexia. In other words, transitions from high school to college and from college to the world of work should be deliberately engineered and coordinated between high schools, colleges and industry to ensure maximum success.

In 2012 Landmark College received approval from our accrediting body, NEASC, to offer 4 year degrees to students with LD. We run various summer term short term programs for high school and college students. Also, Landmark College has initiated a program where we provide college courses (engineered for students who learn differently) online to high school students with LD (including dyslexia), for advanced placement credit and with a high school instructor/facilitator on the ground to assist the student.

Are current educational programs adequate to help those with Dyslexia? Programs specifically designed for students described by a diagnostic label can be short-sighted and stigmatizing. Under the current model, college students first have to establish their eligibility as an individual with a disability under the ADA, and then prove their need for specific accommodations and support services. This is a gatekeeper model that unfairly penalizes students who cannot afford expensive diagnostic evaluations and/or are not well-versed in ways of self-advocacy. However, the amendments to the ADA and recent lawsuits (e.g. the LSAC consent decree) have shifted traditional mindsets. Students no longer have the sole responsibility of having to prove their need for accommodations; the college must should the responsibility of proving why a certain accommodation is service is being denied. This is encouraging.

Educational programs guided by the principles of universal design are both non-stigmatizing and effective because they focus on design and the space, and ways to make the environment inclusive from the onset, rather than having to retrofit for accommodations. UD emphasizes solutions rather than focusing on individual deficits to be remediated or accommodated.

Technology is a vehicle for the application of UD principles within instruction. But care must be taken to ensure that the use of technology does not add to the digital divide (reality that not all have access to technology, internet and similar). Educational programs most beneficial for students with dyslexia should be built around ubiquitous, off-the-shelf, low cost-no cost technologies.

III. **Landmark College Model for Bright College Students with LD.**
Landmark College offers 2- and 4- year degree programs, serves about 500 mostly residential students, and enrolls only neuro-diverse students with an LD such as dyslexia, ADD and/or ASD. Our program is engineered to provide skills and strategies to students who learn differently, and in so doing empower and embolden bright young learners who do not perform well in the mostly one-size-fits-all post-secondary environment. Our graduates (all 2-year; the 4-year BA started two years ago and first graduates will not emerge until 2016) ultimately graduate from baccalaureate programs at a rate higher than all learners (nationally) and well above the graduation rate for college students with LD.

Traditional assessment practices which require diagnostic evaluations can become gate-keepers for students who learn differently. At Landmark College, we take a holistic approach to identification and assessment that seeks learning solutions rather than eligibility and accommodations. While students do need to provide documentation of their disability to enroll at Landmark College, the documentation is one of multiple sources of information that help identify courses that best match students' needs and interests.

In the area of assessment, LCIRT recently hosted a focus group of nationally reputed external evaluators of LD, including diagnosticians from Yale and Harvard University. One of the recommendations of this focus group was to add functional assessments to the traditional portfolio of assessment practices for dyslexia. At Landmark College, our intake process includes many of the recommendations of this focus group.

What is the evidence that our model works? In terms of outcomes and efficacy, our strong retention and graduation data has been mentioned above. In particular, the fact that our graduates demonstrate a higher baccalaureate graduation rate as compared to all (national) learners, let alone those with LD (whom have a low graduation rate), provides empirical proof that our approaches build skills, strategies and confidence that students then employ later in life.

The evidence base for UD in particular, which is our pedagogical under-girding, is quite substantive. References include writings by CAST (Center for Applied Special Technology) David Rose and Anne Meyers. Evidence-based comes in many forms; and testimonials from our students and parents provide yet another form of evidence. Also, recent studies such as one with a Lehigh University collaborator and which looks at archival data from 2006 to 2011 clearly attests to the benefits of our particular brand of student support services (advising and coaching; not yet published).

Landmark College uses a deliberate approach to remediating reading deficits. We provide 1) integrated "assistive" technology that removes reading fluency as the issue, and concentrates on offering active reading, vocabulary development, and visual/auditory interaction to reinforce sound/ symbol relationships, and build comprehension skills. Also, 2) the College offers a Wilson reading program that works with students on phonological awareness, word attack skills, and reading rate and fluency. Finally, 3) the College does not disadvantage students who are poor readers from the opportunity of interacting with college material. Weak readers are not necessarily weak thinkers.

Early intervention with reading is optimal, but schools would do well to approach remediating reading problems by more than skill and drill exercises. Reading in context, providing material appropriate to age level, teaching techniques of skimming, active reading, and monitoring of attention are all important to improving reading ability.

How does Landmark College identify and assess students with dyslexia and other LD? Landmark College requires all students to provide a documented learning disability diagnosis with an updated report written within a three year period. The Academic Affairs Placement Team reads psycho-educational evaluation which includes cognitive scores, achievement testing, diagnostic information and recommendations from a professional clinician.

The diagnosis or learning profile and history of support services (IEP) or accommodations (504). Cognitive ability: IQ is considered as well as processing speed and working memory. Students must have cognitive ability to engage in a college curriculum and work towards the AA or BA/BS degrees. Level of skills: Reading comprehension, fluency, decoding, reading rate, written expression, writing fluency, writing mechanics. The team uses achievement testing and diagnostic criteria to determine placement in one of our three "points of entry" curriculum: Credit, Partial-Credit (hybrid semester) and Language Intensive Curriculum (remedial semester).

The three points of entry. We know that reading issues are not always caused by reading disorders such as dyslexia; attentional learning issues and weak executive functioning can also impact the efficiency and fluency of reading. For this reason we look first at skill level, testing scores and reports from previous school before looking at the diagnosis. Emphasis is placed on learning needs rather than labels.

Students with most severe forms of dyslexia place in a Language Intensive Curriculum. Interventions include combination of remediation and accommodation. Emphasis is on strategy instruction, use of assistive/adaptive technology and the nationally recognized Wilson Reading System. Goal is to access college level material through active learning and use of technology. Students who read and write moderately below college level are placed in Partial Credit Curriculum. Weak vocabulary, lack of background knowledge and dysfluent reading are characteristics. Interventions include developmental writing, reading lab to learn assistive technology (text reader) and teacher directed strategy instruction/study skills. Students who read and write at upper high school or college level are placed in the Credit curriculum with emphasis on learning efficiency. Decoding for these students is not an issue that impacts learning. Fluency issues may persist and may impact efficiency of learning. Interventions include strategy instruction in active reading for comprehension, use of text readers, and writing process methods for composition and clarity of writing.

IV. **Integrated Technologies to Support Learning, and Online Teaching and Learning.**
Schools and colleges need to meet today's students where they reside, namely in the digital sphere and the online platform (but also need to be cognizant of any digital divide). We need intervention solutions for the eLearning ecosystem, not simply assistive technologies for those with dyslexia. The research base for online learning and students with LD is lacking. The merits and demerits of social media specifically for students with LD need much more exploration.

Learning Technologies: This current generation of college students is more technologically savvy than any prior generation. [14] Consider: A recent EDUCAUSE (2013) study[15] of 1,082 university students found that 91% of students own smartphones and 37% have tablets; and 82% of tablet owners use these devices for academic purposes. Students use mobile devices, applications *(apps)*, and social media as an extension of their identity, and not simply as tools or accessories in everyday life. For college students with LD, technologies, including mobile devices and apps offer tremendous promise.[16]

Every day on college campuses around the country, students with LD are advised and coached on ways to use "assistive technologies" to accommodate and/or compensate for their academic difficulties[17]. Although this term is used, even by us, we recognize that the emotional overlay of "assistive" technology can be both stigmatizing and burdensome. The new genres of innovative technologies could make "assistive" a thing of the past. A review of blogs, posts, and tweets by college students reveals the multitude of ways in which students are designing and using technology for creativity, communication, critical thinking, and collaboration; the very skills identified by the National Education Association as critical for the our global society. Yet, little is known of mobile device and apps use among college students with LD. More research is needed in this area.

Landmark College also recently received funding from LDFA (Learning Disabilities Foundation of America) to sponsor a competition on innovative applications or existing and/or new idea for iPad apps, created in partnership between students and faculty. This project aims to understand the technologies that college students today are using; and develop pedagogical applications to support the learning needs of students with dyslexia. The project capitalizes on college students' knowledge of and familiarity with iPads and apps, and provides a means for bridging the gap between student use and pedagogical practice.

60

Finally, **online learning is here to stay, and students with LD are disadvantaged**. Landmark College is developing online teaching and learning platforms that are best engineered to students with LD such as dyslexia. We are collaborating with others on adaptive learning software elements and we are working to build UD into online/blended courses. Through our online efforts we will create an infrastructure that will allow us to develop programs at the post-baccalaureate level which can help us "educate the educators" such as our existing professional certificate program in *Universal Design: Technology Integration*. http://www.landmark.edu/academics/degree-and-credit-options/certificate-program

V.   LD and Dyslexia - and Careers/Workforce

Individuals with dyslexia are a huge source of untapped labor in this country. Just to focus on STEM here, there has been a long-standing recognition and support by the US Department of Labor that the U.S. is not training enough STEM workers to fill economic demand. Persons with disabilities, including those with dyslexia, are underrepresented in STEM across educational and workforce settings.[18][19]

There is a need to improve postsecondary educational opportunities for students with Dyslexia who are interested in STEM careers[19]. According to the United States Government Accountability Office (2009) the number of undergraduate students with disabilities has continued to increase over the past decade reaching a total of 11% of the postsecondary student population. Unfortunately, many qualified college-ready students with LD drop out before completing their college degrees.[20]

Special skills and talents that Dyslexics can bring to the work force. An area where individuals with dyslexia often excel is *entrepreneurship*. Famous examples include: Richard Branson, Whoopi Goldberg, Charles Schwab. These are just the most recognizable names; there are many more individuals with dyslexia who bring entrepreneurial initiatives to our economy. However, a large number of entrepreneurial talent goes unnoticed. Entrepreneurship has to be nourished, especially in the early stages through internships, apprenticeships and sponsored opportunities by industry. Unfortunately postsecondary internship opportunities usually go to those who excel on traditional markers of competency, such as grades. Many students with dyslexia may not have the best GPA or academic score card, but given the opportunity to be creative and innovative, can become the next successful entrepeneur.

Workforce training should include instruction in and access to computer skills that augment reading issues and provide compensatory strategies. Instruction that is offered should not rely solely on auditory memory as auditory processing difficulties are often tied to the dyslexic diagnosis.

Internships and other opportunities for applied/experiential learning are important. Not only do these provide students with LD invaluable experience in the actual work environment, but these efforts better expose employers and our citizens to the strengths of people with a "disability" and ipso facto demystify LD and enlighten all involved. Landmark College is actively pursuing internship opportunities with area industry and other colleges. Example: WHEC Internship project at Landmark College which is a collaborative program with 6 higher education institutions in Windham County, VT.

VI.   Research and Development in LD/Dyslexia (and NSF Funding to Landmark College)

Organizations such as NSF have been supporting educational research to improve outcomes for students with LD, including dyslexia, for some time now. NSF REAL (Research in Education and Learning) is now part of the broader umbrella funding all diversity grants including women's studies, minorities etc. While the philosophical intent of including dyslexia under diversity is laudable, funds within this track should be expanded to address this larger agenda.

In addition to the LDFA (and other) funding, LCIRT at Landmark College recently received two NSF grants to address learning needs of students with LD. This is a testament to efforts by the government to support programs that particularly seek to understand students with learning differences. Such effort is to be lauded. NSF should consider working with institutions to support and extend the outcomes of the projects beyond the funding cycle. Recent NSF grant awards to Landmark College:

1. **National Science Foundation Research in Education and Learning – (NSF-REAL)**

    *   **Awarded:** Landmark College (LCIRT)
    *   **Title:** *Social Presence in Instructor Mediated Synchronous Versus Asynchronous On-Line Discussions: A Study of Undergraduate Students with Disabilities Learning Statistics*
    *   **Award duration:** September 1, 2014 – August 31, 2017 (3 years)
    *   **Award amount:** $486,970
    *   **PI:** Dr. Ibrahim Dahlstrom-Hakki   **Co-PI:** Dr. Manju Banerjee

    Project Summary: We have known for some time now that instructor presence is a critical learning need for students with LD. Online learning is a distraction-rich environment which can be particularly challenging for students with LD and executive function difficulties. This project will investigate a critical question about students with LD in online courses, namely, *the importance of instructor-mediated synchronous discussions within an online platform.* For the purposes of this study, instructor-mediated synchronous discussions will be conducted as live virtual group interactions (instructor and 8-12 students) via Adobe Connects. Each participant will be able to see everyone else, share screens, and talk in real-time about the course topic or class assignments. Synchronous discussions will be compared to asynchronous discussions conducted via VoiceThread, where students will share aloud their thinking via video capture online, though not in real time.

    Given the paucity of research on best practices for online innovations, the results of this study from Landmark College will inform teachers/instructor, instructional designers, and institutional perspectives on emerging models of online education for diverse learners.

2. **National Science Foundation Data Intensive Research to Improve STEM teaching and Learning**

    *   **Awarded:** Landmark College (LCiRT); MIT, TERC, Inc.
    *   **Title:** *Revealing the Invisible: Data-Intensive Research Using Cognitive, Psychological, and Physiological Measures to Optimize STEM Learning*
    *   **Award duration:** August 15, 2014 Start – (2 and 1/2 years)
    *   **Award amount:** Total: **$1,163,711 million**; Landmark College award **$270,363**
    *   **PIs:** Dr. Ibrahim Dahlstrom-Hakki (Landmark College); Dr. Jodi Asbell-Clarke (TERC, Inc.); Dr. Micah Altman (MIT)

    Project summary: This project brings together expertise in learning sciences, cognitive psychology and data sciences to advance core knowledge about how big data, enhanced with biometric information, can aid in the study of learning. The goals of this exploratory research are to understand how *exhaust data* [21] from digital games can be used to customize optimal learning experiences; and more broadly, evaluate how exhaust data can reveal basic cognitive activities that are prerequisites for learning. More specifically, the project will study the relationship among constructs of gaming activity such as engagement, attention and memory, by having students play science video games and observing their implicit learning of basic principles of Newtonian physics.

VII.   (nota bene) Vice President Nelson Rockefeller

House members likely know the story of one of their own. Nelson Rockefeller served as the 41st Vice-President of the United States from 1974-1976. He was also the 49th Governor of New York and worked in the administrations of Franklin Roosevelt, Harry Truman, and Dwight Eisenhower. Rockefeller also had dyslexia. Nelson Rockefeller almost single-handedly changed how people thought about dyslexia when, in 1976, he "came out" to the public about his own dyslexia. It was perhaps the first time a major public figure, a Vice-President, and someone from such an illustrious and successful family had openly discussed his dyslexia.

In 1976, while serving as Vice-President, Rockefeller discussed his own severe case of dyslexia in a popular magazine article. He noted how dyslexia impacted his performance in school and also how he had to memorize his speeches during his political career because he was afraid of trying to read them. This was an important moment in the American public's recognition that dyslexia was a common learning disability, affecting even the most prominent families. It also helped the public understand that dyslexia did not impact intelligence, motivation, or character – that even those with dyslexia could become highly successful adults, even rising to the ranks of one heartbeat away from the Presidency.

In the magazine interview, Rockefeller recalled, "I was dyslexic... and I still have a hard time reading today. I remember vividly the pain and mortification I felt as a boy of eight when I was assigned to read a short passage of scripture at a community vesper service and did a thoroughly miserable job of it. I know what a dyslexic child goes through... the frustration of not being able to do what other children do easily, the humiliation of being thought not too bright when such is not the case at all. But, after coping with this problem for more than 60 years, I have a message of hope and encouragement for children with learning disabilities and their parents." [22]

References

1.   Sherman, G. (2012). Neuroscience in the 21st century: Where are we going. Paper presented at the 63rd Annual International Dyslexia Association conference, October 24-27, Baltimore, MD.
2.   United States Government Accountability Office, 2009
3.   Wagner, M., Marder, C., Blackorby, J., Cameto, R., Newman, L., Levine, P., & Davies-Mercier, E. (with Chorost, M., Garza, N., Guzman, A., & Sumi, C.) (2003). *The achievements of youth with disabilities during secondary school. A report from the National Longitudinal Transition Study-2 (NLTS2).* Menlo Park, CA: SRI International. Available at www.nlts2.org
4.   Wagner, M., Newman, L., Cameto, R., Garza, N., & Levine, P. (2005). *After high school: A first look at the postschool experiences of youth with disabilities. A report from the National Longitudinal Transition Study-2 (NLTS2).* Menlo Park, CA: SRI International. Available atwww.nlts2.org
5.   *The State of Learning Disabilities: Facts, Trends and Emerging Issues* Third Ed. 2014 | LD.org

63

House Hearing: Science of Dyslexia   Landmark College; P. Eden, Ph.D.

6. Sanford, C., Newman, L., Wagner, M., Cameto, R., Knokey, A.-M., and Shaver, D. (2011). *The Post-High School Outcomes of Young Adults With Disabilities up to 6 Years After High School. Key Findings From the National Longitudinal Transition Study-2 (NLTS2)* (NCSER 2011-3004). Menlo Park, CA: SRI International.

7. Bureau of Labor and Statistics, U.S. Department of Labor. (2012). News release: Persons with a disability: Labor force characteristics — 2013, Friday, June 11, 2014. Retrieved from: http://www.bls.gov/news.release/pdf/disabl.pdf

8. Cortella, Candace, & Horowitz. (2014). *The state of learning disabilities: Facts, trends, and emerging issues.* New York: National Center for Learning Disabilities

9. H. Stephen Kaye, *Unpublished tabulations of 2005 data from the U.S. Survey of Income and Program Participation* in C. Cortiella, The State of Learning Disabilities 2010. National Center for Learning Disabilities, New York, NY.

10. *The State of Learning Disabilities: Facts, Trends and Emerging Issues* Third Ed. 2014 | LD.org

11. Newman, L., Wagner, M., Knokey, A.M., Marder, C., Nagle, K., Shaver, D., Wei., X, Cameto, R., Contreras, E., Ferguson, K., Greene, S., & Schwarting, M. (2011). *The post-high school outcomes of young adults with disabilities up to 8 years after high school: A report from the National Longitudinal Transition Study—2 (NLTS2)* (NCSER 2011-3005). Menlo Park, CA: SRI International

12. U.S. Department of Education, Institute of Education Sciences, National Center for Special Education Research, National Longitudinal Transition Study-2 (NLTS2), Wave 1 school program survey, 2002; responses restricted to those who eventually enrolled in postsecondary schools and had disclosed a disability to their postsecondary school

13. *A Review of Postsecondary Programs Designed for Students with Learning Disabilities, Attention Deficit Hyperactivity Disorder, or Autism Spectrum Disorder.* Report prepared for Landmark College by Hanover Research, February, 2012.

14. Dahlstrom, Eden, J.D. Walker, & Charles Dziuban (2013). ECAR Study of Undergraduate Students and Information Technology, (Research Report). Louisville, CO: EDUCAUSE Center for Analysis and Research, Retrieved March 5, 2014 from https://net.educause.edu/ir/library/pdf/ERS1302/ERS1302.pdf

15. Chen,B. & Denoyelles, A. (2013). Exploring students' mobile learning practices in higher education. Retrieved March 2014 from http://www.educause.edu/ero/article/exploring-students-mobile-learning-practices-higher-education [This is the EDUCAUSE 2013 study]

16. Smith, S. J., & Okolo, C. (2010). Response to intervention and evidence-based practices: Where does technology fit? *Learning Disability Quarterly, 33,* 257-272.

17. Lindstrom, J. H. (2007). Determining appropriate accommodations for postsecondary students with reading and written expression disorders. *Learning Disabilities Research & Practice, 22*(4), 229-236.

18. Burrelli, J. S., & Falkenheim, J. C. (2011). Diversity in the federal science and engineering workforce. Washington, DC: National Science Foundation, Directorate for the Social, Behavioral, and Economic Sciences (NSF 11-303).

19. National Science Foundation. (2009). Women, minorities, and persons with disabilities in science and engineering. Washington, D.C.: National Science Foundation (NSF 09-305). National Science Foundation, National Center for Science and Engineering Statistics. 2013. *Women, Minorities, and Persons with Disabilities in Science and Engineering: 2013.* Special Report NSF 13-304. Arlington, VA. Retrieved September 2014 from http://www.nsf.gov/statistics/wmpd/.

20. Stamp, L., Banerjee, M. & Brown, F. (2014). Self-advocacy and perceptions of college readiness among students with AD/HD. *Journal of Postsecondary Education and Disability, 27*(2), 12-47

64

21. Halverson & Owen, (2013) Halverson, R., Wills, N. & Owen, E. (2012). CyberSTEM: Game-based learning telemetry model for assessment. Paper presented at the 8th annual Games and Learning and Society (GLS). conference 8.0, Madison, WI.

22. Source: Rockefeller, Nelson (1976, October 16). *TV Guide*, pp. 12-14 (pp.12-14).

Dyslexia and LD in Post-Secondary Education - Challenges and Opportunities

P. Eden, Ph.D. Landmark College, Putney VT

Students with Dyslexia are Disadvantaged

- Hundreds of thousands in colleges and universities (many more with other LD)

- Most in two year colleges. Most (90%) receive accommodations in high school but majority (83%) do NOT in college

- BA/BS graduation rates for students with LD/dyslexia low (about 37%)

- Rarely find comprehensive support programs (let alone innovative ones….)

- Online teaching/learning: Standard modalities will exacerbate challenges for LD/dyslexia

2

Innovative Educational Practices and Efforts

- Universal Design (UD/UDL): Engineer learning environment to anticipate neurodiversity, heterogeneity of students and learning profiles. Multiple modes of presentation, responses, engagement. Conventional or online classes.

- Mobile Technologies: Meet students where they exist. Develop "assistive" and integrated technologies on hand-held devices (ubiquitous, not specialized rooms/tools)

- Cognitive Training: Explore patterns of learning through markers (cognitive, physiological etc.) via video game activities. Analyze big data and exhaust data…seek adaptive and customizable learning activities.

Landmark College Model

- Offer 2- and 4-year degree programs (including STEM)

- Serve ONLY students with LD such as dyslexia, or ADD and/or ASD. Dedicated; not add-on resource or program

- Use UD/UDL and integrated technologies, careful placement in curriculum tracks, hidden curriculum of support

- Retention rates high. Ultimate BA/BS graduation rate (70%) higher than national average (59%) for ALL learners.

- Juxtapose research and innovation with teaching and learning

4

Research and Development at Landmark College

- Landmark College Institute of Research and Training (LCIRT)

- Recent funding (LDFA) for iPad APP (learning) development by students and faculty. Also, recent NSF funding:

- NSF-REAL (Research in Education and Learning): Instructor presence in synchronous elements for online learning; STEM content and students with LD

- NSF Data Intensive Research to Improve STEM Teaching and Learning (MIT collaboration): Revealing the invisible; Gaming vehicle, engagement, eye-tracking, attention, memory...implicit understanding of principles of Newtonian physics. Big data (n), exhaust data analysis...

5

Dyslexia and LD in Post-Secondary Education Summary

- LD/dyslexia barriers to learning, education, employment

- Huge, untapped population of potential skilled workers (e.g. in STEM)

- Innovative educational practices (scalable) needed, particularly with online education realities

- Use technology to discover better teaching and learning platforms and understand neurodiverse students - and to provide ubiquitous tools for success in college and careers

- Research and support increasingly focused in this area (more needed)

6

Peter A. Eden – Biography

Peter A. Eden, Ph.D., took office as the fourth president of Landmark College on July 1, 2011. A scientist and researcher, he also is an accomplished educator and administrator whose passion for learning, innovation and discovery infuses his aspirations for the College's future.

A firm believer that "education is ripe for reinvention," Dr. Eden embraces technology and champions the responsible integration of new technology in higher education. With a strong background in modern science, research, teaching and scholarship, he stands poised to lead Landmark College into a new era of learning for college students with learning disabilities, ADHD, and autism spectrum disorder (ASD).

Dr. Eden's varied experience encompasses nearly 20 years of achievement as an administrator and teacher in higher education, as well as a researcher and project director in the biotech/pharmaceutical industry.

Prior to joining Landmark, Dr. Eden served as dean of arts and sciences and professor of biotechnology at Endicott College in Beverly, MA. He previously was a tenured associate professor and chair of the science department at Marywood University in Scranton, PA. He was also a research fellow at the Jackson Laboratory and a visiting professor at the College of the Atlantic, both in Bar Harbor, ME. Before that, he worked five years at Biomeasure, Inc. (Beaufour-IPSEN) in Milford, MA, and Paris, France — initially as a molecular biologist, then quickly advancing to research project director.

Dr. Eden earned his undergraduate degree at the University of Massachusetts Amherst. He completed his Ph.D. at the University of New Hampshire and his post-doctoral training at Massachusetts Institute of Technology in microbiology, molecular biology, and neurobiology.

He has published more than 20 scientific articles, received NIH and NSF grant funding for research, and led the development of innovative undergraduate and graduate academic programs.

Forward thinking and progressive, Dr. Eden is exploring the development of new programs that promise to enhance learning at Landmark and ensure that the College remains the leader in higher education for students with learning disabilities, ADHD,and ASD.

Dr. Eden and his wife, Joanne, are the parents of three young children, Atticus, Dexter, and Vivienne.

Chairman SMITH. Thank you, Dr. Eden. We will go to the next Dr. Eden, though no relation.

**TESTIMONY OF DR. GUINEVERE EDEN,
DIRECTOR, CENTER FOR THE STUDY OF LEARNING (CSL)
AND PROFESSOR, DEPARTMENT OF PEDIATRICS,
GEORGETOWN UNIVERSITY MEDICAL CENTER**

Dr. GUINEVERE EDEN. Thank you, Chairman Smith, for holding this hearing and for the invitation to speak to you today about the brain-based scientific understanding of dyslexia.

Magnetic resonance imaging, as you have already heard, has provided a way by which researchers can study the brain and anatomy and function noninvasively, thereby permitting the study of children. Since my colleagues and I first implemented functional magnetic resonance imaging to study dyslexia in 1996, the understanding of dyslexia has advanced significantly.

Reading stands out in cognitive neuroscience. It is a uniquely human skill and cannot be ecologically simulated in animal models. At Georgetown University, we use functional MRI to study the reading brain in action, the developmental trajectory of reading, the difference in people with dyslexia. We have examined skills other than reading noted to be affected in people with dyslexia to evaluate which underlying brain differences are causal to the reading problems and which are not. We have begun to investigate males and females separately since our findings in girls and women suggest that the brain mechanisms for dyslexia may in part be sex-specific. We have examined the impact of intensive reading intervention and learned that adults with dyslexia not only make gains in reading but exhibit plasticity as demonstrated by increased brain activity. Intervention also results in growth in brain tissue. As such, reading gains in dyslexia are brought about by complex physiological and anatomical changes.

The main challenge for our research is determining the etiology of these brain-based findings, and researchers across the country are tackling this very question. Molecular mechanisms have been probed by examining MRI scans in children with dyslexia for the chemicals that support communication amongst brain cells. Also, studies have been conducted in people who are carriers of dyslexia-associated genes to better understand the gene-brain relationship. Here, animal models have been very useful. Mice specifically bred to carry dyslexia-associated genes are studied to determine how these genes operate at the cellular level, thereby filling the void where human research is limited. Together these NIH-supported studies have improved our understanding and raised awareness for the complexity of dyslexia.

How can these findings be translated? We now know that learning to read, as my first-grade daughter is doing at this very moment, eventually leads to substantial changes in brain anatomy and brain function. Will brain imaging allow us to identify dyslexia in pre-readers or forecast who might benefit from intervention, and of what kind? Neuroscientists are working on these possibilities, and imaging data are proving indicative of future reading outcome in dyslexia.

Factors constraining these efforts are mostly technical in nature and could be surmounted by technological advances such as those envisioned in the President's BRAIN Initiative, potentially allowing for observations at the individual as opposed to the group level, and in younger children.

A continuing barrier in the field is the distance between academic research and educational practices. Researchers publish in specialty journals, which are often inaccessible to those who operate as educators in the field. Teachers may therefore not be implementing the approaches that have been proven to be successful by rigorous research studies. Conversely, researchers are at risk of pursuing theories that are not relevant to real classroom settings.

Some agencies have addressed this problem. The National Science Foundation Science of Learning Centers, for example, like the one here at Gallaudet University allow for an environment to integrate knowledge across multiple disciplines and connect research with educational challenges. These conduits need to be increased if we are to move—have more than a dialog spanning the gamut from neuroscience to classroom activities. More training opportunities that expand the knowledge base in each field with respect to the other and funding opportunities that promote collaboration are needed.

In the meantime, others are stepping up to fill the gap. For example, the International Dyslexia Association, the IDA, has provided guidelines for the training of teachers of students with dyslexia based on current research. Further, the IDA is providing accreditation to those universities engaged in teacher training that abide by these high standards, allowing for those teachers to have the necessary skills to identify and teach children with dyslexia effectively. This is in addition to the longstanding efforts by the IDA to bring researchers, practitioners and parents together to provide and share information and resources, and in ways that are relevant and accessible to each stakeholder. The IDA and other nonprofit organizations raise awareness and distribute knowledge. This can also protect parents and educators from seemingly promising commercial programs for dyslexia that in reality provide little or no benefit.

Overall, the science of dyslexia has made significant advances. Challenges have arisen, which can be met by Federal support for science and education intertwined, allowing changes in academic and educational institutions that will facilitate jointly tackling the collective complexity of dyslexia and harness the knowledge of teachers and science of learning to the benefit of people with dyslexia.

Thank you.

[The prepared statement of Dr. Eden follows:]

Testimony before the U.S. House of Representatives Committee on Science, Space, and Technology

2321 Rayburn House Office Building

Washington DC, 20515

The Science of Dyslexia

September 18th, 2014

Guinevere F. Eden, D.Phil.

Professor of Pediatrics

Director, Center for the Study of Learning,

Georgetown University Medical Center,

Washington, DC 20057

edeng@georgetown.edu

Thank you, Chairman Smith, for holding this hearing and for the invitation to speak to you today about the brain-based scientific understanding of dyslexia.

Brain Imaging Technology has Advanced Our Understanding of the Brain Bases for Reading and Dyslexia

Since the 1991 US-led invention [1] of functional magnetic resonance imaging (fMRI), there has been an explosion in the use of this technique for the purpose of observing the locations and characteristics of activity in the human brain underlying sensation and cognition. While researchers had already been using standard MRI to scrutinize brain structural differences in dyslexia, fMRI has allowed researchers to visualize the reading brain in action. Unlike other areas of cognition, reading is a uniquely human skill and cannot be ecologically simulated using animal models. Further, the non-invasive nature of this technique allows for tracking children over time. Since our first implementation of fMRI to study dyslexia at the National Institutes of Health in 1996 [2], the field has grown rapidly and made significant contributions to the science of dyslexia.

Reading, a cultural invention that allows us to represent speech as symbolic form, involves a coordination of the brain's language areas with the visual and auditory systems. At my center at Georgetown University, we have studied brain activity while participants process words [3]. We use this approach to characterize the developmental trajectory of reading acquisition [4], study non-alphabetic reading [5], and uncover differences in people with dyslexia [6]. To address the multitude of theories that have been proposed to explain dyslexia, we have also studied the brain as it engages in other tasks thought to be affected by reading disability [7]. We have examined the impact of intensive reading intervention and learned that adults with dyslexia not only make gains in reading, but also show brain plasticity, as demonstrated by increases in brain activity [6]. Brain anatomy is also malleable: in another study we found that reading intervention resulted in growth of brain tissue [8]; together, these studies illustrate how reading gains in people with dyslexia are brought about by complex physiological and anatomical brain changes.

We sometimes encounter brain-based observations for which there were no obvious indications from behavioral studies. For example, my research group has found that the brains of females with dyslexia do not conform to the neurobiological model of dyslexia that was largely derived from studies of males [9]. This might have important implications for diagnosing and treating females with dyslexia.

Some of the same brain areas that are compromised for reading are also underactive when children with dyslexia solve arithmetic tasks [10], highlighting the far-reaching consequences of dyslexia and their complex connection to other forms of learning disabilities.

What remains challenging for much of this research is to be able to assess what is directly causing the reading problems and distinguish these factors from those that are a consequence or a byproduct of whatever is causing the dyslexia [11].

There have been notable discoveries across the country that demonstrate the interdisciplinary nature of this work. Linking with investigations into the genetic mechanisms of dyslexia, brain imaging studies have been conducted in carriers of dyslexia-associated genes [12]. Here is where animal models can and have been strategically employed: knock-out mice are used to find out how these dyslexia-associated genes operate ([13,14] Harvard University and University of Connecticut), filling the void where human research is limited. The molecular mechanisms have also been probed by examining MRI scans of children with dyslexia for the brain chemicals that support communication amongst brain cells ([15,16] Haskins Laboratory, CT, and University of Southern California, CA). Together these studies, which are supported by the NIH, have improved our understanding and raised awareness of the complexity of dyslexia.

What are the Possibilities and What are the Limitations?
Imaging technology facilitates characterization of the intricate developmental changes that occur in our children's brains and the formal learning they experience in our educational system. We now know that learning to read–as my first-grade daughter is doing at this very moment–eventually leads to substantial changes in brain anatomy and function. Will brain imaging allow us to identify dyslexia earlier than first grade, or forecast who might benefit from intervention and of what kind? Neuroscientists are working on these possibilities, and imaging data are proving indicative of future reading

outcomes in dyslexia ([17] Stanford University, CA). Factors constraining these efforts are mostly technical in nature. <u>Future technological advances may allow us to surmount these hurdles</u>. The development of better technologies, as envisioned in the President's BRAIN initiative, should continue to improve methods of monitoring the human brain, allow for observations that are based on one individual person rather than a group of people, with application to younger children, and with the ability integrate information across different levels of inquiry.

Practical Implications: How is the Knowledge Applied?

Academic researchers are bound by academic practices to publish in specialty journals. These are often inaccessible, physically and conceptually, to those who directly operate as educators in the field. Researchers are at risk of pursuing theories that are not relevant to real classroom settings. Conversely, teachers may not be implementing approaches that have been proven to be successful by rigorous research studies. As such, a significant barrier is the physical and cultural distance between academic research and educational practices.

Some agencies have addressed this problem via targeted funding mechanisms. The NSF Science of Learning Centers are a notable example of creating an environment to integrate knowledge across multiple disciplines, establishing common grounds for conceptualization and connecting research with educational challenges. These conduits need to be increased if we are to have more than a dialogue spanning the gamut from neuroscience to classroom activities.

Others are stepping up to fill the gap. For example, to address concerns that basic research about reading is not available to teachers, the International Dyslexia Association (IDA) has provided guidelines on the desired capabilities for teachers of students with dyslexia ("Knowledge and Practice Standards for Teachers of Reading" [18]). These have also been used by the IDA to accredit those university teacher training programs in reading that promote high standards for comprehensive and rigorous training of teachers [19]. This is in addition to the longstanding efforts by the IDA to bring researchers, practitioners, and parents together (for example, through their international and local conferences) and to provide resources by which parents can learn about dyslexia, gauge the current state of research and practice, and make decisions about their child with information that is relevant and accessible to them.

Less welcome contributors to this arena are commercial entities that purport to marry research-based knowledge to address educational needs, such as poor reading skills, but use questionable approaches and put the goal of creating profit before the goal of translating research into educational gains for children. These need to be countered by efforts in which researchers and educators work together. These problems can be addressed by training opportunities that expand the knowledge base in each field with respect to the other, and by funding opportunities that promote collaboration.

Overall, the science of dyslexia has made significant advances in the last 25 years. With these, challenges have arisen which can be met by federal support for science and education intertwined, by academic and educational institutions embracing a cultural change that facilitates jointly tackling the collective complexity of dyslexia, and engaging a common language and a common understanding of how to harness the knowledge of teaching and learning to the benefit of children with dyslexia.

References

[1]   J.W. Belliveau, D.N. Kennedy, R.C. McKinstry, B.R. Buchbinder, R.M. Weisskoff, M.S. Cohen, et al., Functional mapping of the human visual cortex by magnetic resonance imaging, Science. 254 (1991) 716–719.

[2]   G.F. Eden, J.W. VanMeter, J.M. Rumsey, J.M. Maisog, R.P. Woods, T.A. Zeffiro, Abnormal processing of visual motion in dyslexia revealed by functional brain imaging, Nature. 382 (1996) 66–69. doi:10.1038/382066a0.

[3]   O.A. Olulade, D.L. Flowers, E.M. Napoliello, G.F. Eden, Developmental differences for word processing in the ventral stream, Brain Lang. 125 (2013) 134–145. doi:10.1016/j.bandl.2012.04.003. PMCID: PMC3426643.

[4]   P.E. Turkeltaub, L. Gareau, D.L. Flowers, T.A. Zeffiro, G.F. Eden, Development of neural mechanisms for reading, Nat. Neurosci. 6 (2003) 767–773. doi:10.1038/nn1065.

[5]   Krafnick, A.J., Luetje, M., Napoliello, E.M., Flowers, D.L., Tan, L.H., and Eden, G.F., English word, Chinese character and object processing in young English speaking children, Organ. Hum. Brain Mapp. Abstr. 7323. (2012).

[6]   G.F. Eden, K.M. Jones, K. Cappell, L. Gareau, F.B. Wood, T.A. Zeffiro, et al., Neural changes following remediation in adult developmental dyslexia, Neuron. 44 (2004) 411–422. doi:10.1016/j.neuron.2004.10.019.

[7]   O.A. Olulade, E.M. Napoliello, G.F. Eden, Abnormal visual motion processing is not a cause of dyslexia, Neuron. 79 (2013) 180–190. doi:10.1016/j.neuron.2013.05.002. PMCID: PMC3713164.

[8]   A.J. Krafnick, D.L. Flowers, E.M. Napoliello, G.F. Eden, Gray matter volume changes following reading intervention in dyslexic children, NeuroImage. 57 (2011) 733–741. doi:10.1016/j.neuroimage.2010.10.062. PMCID: PMC3073149.

[9]   T.M. Evans, D.L. Flowers, E.M. Napoliello, G.F. Eden, Sex-specific gray matter volume differences in females with developmental dyslexia, Brain Struct. Funct. 219 (2014) 1041–1054. doi:10.1007/s00429-013-0552-4. PMCID: PMC3775969.

[10]   T.M. Evans, D.L. Flowers, E.M. Napoliello, O.A. Olulade, G.F. Eden, The functional anatomy of single-digit arithmetic in children with developmental dyslexia, Neuroimage. (2014).

[11]   A.J. Krafnick, D.L. Flowers, M.M. Luetje, E.M. Napoliello, G.F. Eden, An investigation into the origin of anatomical differences in dyslexia, J. Neurosci. 34 (2014) 901–908. doi:10.1523/JNEUROSCI.2092-13.2013. PMCID: PMC3891966.

[12]   N. Cope, J.D. Eicher, H. Meng, C.J. Gibson, K. Hager, C. Lacadie, et al., Variants in the DYX2 locus are associated with altered brain activation in reading-related brain regions in subjects with reading disability, NeuroImage. 63 (2012) 148–156. doi:10.1016/j.neuroimage.2012.06.037.

[13]   A.M. Galaburda, J. LoTurco, F. Ramus, R.H. Fitch, G.D. Rosen, From genes to behavior in developmental dyslexia, Nat. Neurosci. 9 (2006) 1213–1217. doi:10.1038/nn1772.

[14]   A. Che, M.J. Girgenti, J. Loturco, The Dyslexia-Associated Gene Dcdc2 Is Required for Spike-Timing Precision in Mouse Neocortex, Biol. Psychiatry. in press (2014). doi:10.1016/j.biopsych.2013.08.018.

[15]   K.R. Pugh, S.J. Frost, D.L. Rothman, F. Hoeft, S.N.D. Tufo, G.F. Mason, et al., Glutamate and Choline Levels Predict Individual Differences in Reading Ability in

Emergent Readers, J. Neurosci. 34 (2014) 4082–4089. doi:10.1523/JNEUROSCI.3907-13.2014.

[16] J.L. Bruno, Z.-L. Lu, F.R. Manis, Phonological processing is uniquely associated with neuro-metabolic concentration, NeuroImage. 67 (2013) 175–181. doi:10.1016/j.neuroimage.2012.10.092.

[17] F. Hoeft, B.D. McCandliss, J.M. Black, A. Gantman, N. Zakerani, C. Hulme, et al., Neural systems predicting long-term outcome in dyslexia., PNAS. 108 (2011) 361–6. doi:10.1073/pnas.1008950108.

[18] International Dyslexia Association, Professional Standards and Practices Committee, Knowledge and Practice Standards for Teachers of Reading, (2010). http://www.interdys.org/ewebeditpro5/upload/KPSJul2013.pdf.

[19] International Dyslexia Association, University Programs Accredited by IDA, (2014). http://www.interdys.org/AccreditedUniversityPrograms.htm.

Guinevere Eden is a Professor in the Department of Pediatrics and the Director of the Center for the Study of Learning at Georgetown University Medical Center. Her research has been supported by grants from the National Institutes of Health and the National Science Foundation.

Dr. Eden and her colleagues were the first to apply functional MRI to the study of dyslexia in 1996, and she continues to investigate the neural bases of dyslexia and the neural correlates of successful reading intervention.

She is past-president of the International Dyslexia Association, a non-profit organization dedicated to helping individuals with dyslexia, their families, and the communities that support them by promoting literacy through research, education, and advocacy. She has also served as Scientific Co-Director for the National Science Foundation-funded "Science of Learning Center" housed at Gallaudet University.

Dr. Eden received her B.S. in Physiology from University College London, her Ph.D. in Physiology from the University of Oxford, and conducted her postdoctoral training at the National Institutes of Health.

Dr. Eden has published widely (including journals such as *Nature, Nature Neuroscience,* and *Neuron*) and is a frequent speaker in the US and internationally. She has served as a permanent member of a standing NIH Study Section and as chair for several Special Emphasis Panels. She serves on the Editorial Boards for the journals *Annals of Dyslexia, Dyslexia, Brain and Language, Developmental Cognitive Neuroscience,* and *Human Brain Mapping.*

Chairman SMITH. Thank you, Dr. Eden, and I will recognize myself for questions, and each Member is limited to five minutes, so I am going to ask brief questions, and Dr. Shaywitz, let me address the first one to you.

You noted, and we are aware, that various progress has been made in our understanding of dyslexia, but what would be some next steps that we need to take to better that understanding? Could you put your microphone on there?

Dr. SHAYWITZ. There are a number of next steps, but my emphasis is that we know so much, and I worry that people will say well, wait until we find out more and more. Children have one life to live. We have to make sure that we address their needs now; we have more than enough knowledge to make changes. We have heard from Max Brooks and Stacy Antie about what dyslexia does to a child and to a family, and how can we let that continue to go on when we have the knowledge to make a difference?

Chairman SMITH. So use the knowledge we have better than we do?

Dr. SHAYWITZ. Use it, implement it, have teachers and parents who are aware of the signs of dyslexia and what are evidence-based programs that will help the children. They can't hold their breaths waiting—we always need more knowledge but dyslexia is in the unique position, we have so much knowledge. A parent goes to a school and says, "we don't believe in dyslexia." That is unacceptable.

Chairman SMITH. Okay. Thank you, Dr. Shaywitz.

Dr. SHAYWITZ. You are welcome.

Chairman SMITH. Mr. Brooks, you ought to be a talk-show host as long as you didn't mention Congress. Don't answer that.

The question is this. How has dyslexia helped you in your career?

Mr. BROOKS. Well, I think it has helped me mentally and emotionally. I think, as I said before, it helped me mentally because I couldn't simply regurgitate facts. I had to make sure that I understood the facts, and the best way to understand facts is to understand the bigger picture, and I think it is that big-picture thinking that has helped me.

But also emotionally, it has made me resilient. You know, I think as we all know, one of the biggest challenges to any human being is to get out of your comfort zone. Well, anybody with dyslexia knows you don't have a comfort zone. You are always struggling, so it makes you comfortable with struggling. So in that way, it has made me very comfortable with new challenges.

Chairman SMITH. Okay. Good. Thank you, Mr. Brooks.

Ms. Antie, what should we do in our public schools to help dyslexic kids?

Ms. ANTIE. The way that my son has been taught through the systemic evidence-based curriculum has made the difference in the world to him. It has taught him actually how to decode words, and I think that is beneficial to everybody, because as he taught me, I was able as adult even to say that really does make sense. So I think every child can benefit from having that, and smaller schools—smaller classroom sizes. I know that is not always the best way because, you know, we have so many children in school but 32 down to 16 makes a tremendous difference in the classroom.

Chairman SMITH. Okay. Thank you, Ms. Antie.

Dr. Eden, I really don't have a question for you because you answered my question. Thank you for your suggestions as to what we should do on the college level to help dyslexic students. I appreciate that very much.

Dr. Eden, what should we do to help make reading a priority when we are dealing with dyslexia?

Dr. GUINEVERE EDEN. Well, I think we need to recognize the importance of learning and to recognize that the gateway to learning is learning to read, and so in the absence of reading, it is not just learning to read directly but it is what you learn once you are able to read that is so important. It has to be a priority. It seems so obvious. We are spending so much time, we understand so much about the science of learning, we understand so much about dyslexia, we understand how you can teach students with dyslexia, but for some reason we are just not able to get the information out there and bring it to the teachers who are performing this very important task of teaching our children how to learn to read.

Chairman SMITH. Okay. Thank you, Dr. Eden.

That concludes my questions, and the gentlewoman from Texas, Ms. Johnson, is recognized for hers.

Ms. JOHNSON. Thank you very much.

Dr. Peter Eden, in your testimony you mentioned how Landmark College recently received two NSF grants to address the learning needs of students with learning disabilities like dyslexia. It has been a while since we have had the opportunity to hear about the educational research that NSF is funding, and I am excited that we have the opportunity today. Would you please elaborate on those research grants that Landmark College has received and how that research has led to any type of results for the individuals?

Dr. PETER EDEN. Absolutely. These particular grant proposals were awarded just within the past few weeks, so we don't have the data yet. However, it is very, very exciting to embark on. Both are at least three years in duration. The first NSF award, the REAL, Research in Education and Learning, it focuses on—in an online learning environment with the topic of statistics in the STEM field. Students with LD including dyslexia, will they benefit from instructor presence in a synchronous, not an asynchronous fashion—synchronous means in real time right there—and whether or not that makes a measureable difference in their understanding of the content and the outcomes in this class if in the online environment there is an instructor present to help them as they move through the course, and this will be invaluable because you see the proliferation of online courses for all learners in understanding the efficacy of online STEM courses, whether or not we have a synchronous element with response immediacy and an instructor right there to help them with the content or not, and that will be determined through this grant proposal.

The other is the proposal in collaboration with MIT and TERC, and this is the effort to leverage the fact that we have so many college students, and younger, of course, gaming, playing videogames, and hopefully understanding something about their memory and their attention and even using eye tracking as they understand in this case Newtonian physics built into this science videogame, and when I say big data and exhaust data, that means we could possibly if we have enough remote webcam technology in the future see what thousands, hundreds of thousands perhaps students experiencing these videogames, playing these games, how they learn, how they understand the principles within the videogame and harnessing that big data. So this platform of gaming provides a tremendous window into how students may learn, and we are trying to understand that better through that NSF grant proposal.

Ms. JOHNSON. Thank you very much.

And this question probably will be my final one because of time. To any of the panel members, we have done some research but it seems to me, and of course the type that you are doing, I think is going to be extremely good to have the outcomes, are we getting the benefit of it with people who are working directly and how can we better enhance that? I ask that question because many of you have mentioned going through a struggle to find some answers. Then I hear that many of the answers sometimes can be found through technology rather than making a student read. I just want to get some feedback of what you think needs to be done to better get that information out.

Mr. BROOKS. I can tell you one thing that we can do right now, which is make mandatory dyslexia recognition training part of any teacher's certificate because, look, we make our policemen take mandatory racial sensitivity training. Why not make our teachers take mandatory dyslexic recognition training so maybe they can recognize that the class clown or the troublemaker or the kid who stares out the window is actually compensating for dyslexia. We have all these new solutions and all these solutions that keep coming but none of that is going to help if we don't get the kids to the solutions in the first place, so that is the first step.

Ms. JOHNSON. Thank you. My time is expired, but anybody else can comment.

Chairman SMITH. Normally we don't encourage audience participation, but today it is appreciated.

The gentleman from Indiana, Dr. Bucshon, is recognized.

Mr. BUCSHON. Thank you, Mr. Chairman.

My daughter does not have dyslexia but she did have a learning disability. We started to notice really in kindergarten that she was not able to read as well as her older brothers, and in first grade she started to get behind, way behind with a different type of learning disability. And to your point, Mr. Brooks, this became a social problem for my daughter, and she came to her parents wondering if she was dumb. So this is a big deal, and my point is this: Parents have to advocate early. The system does want to help but I always think all the time if she didn't have parents there advocating right then—we got her tested. We found out what her problem was. She got specific direction. Again, she does not have dyslexia but she had a short-term memory thing. It is like tell her seven items in a row. Most kids her age could regurgitate five or six. She could only remember two. I don't know what that is called but basically that was her problem, and she has come around. That is ten years ago now. She is a straight A student in high school. So thank you for your testimony about the social aspects of this type of problem. It is real. She had—and it took her really years to recover the self-esteem but now she has, so that is very important.

My question I think has already been answered but I will ask Dr. Shaywitz. The National Center for Learning Disabilities believes that all parents and professionals who work with young children should be informed the early signs of dyslexia. I guess maybe you can focus on what families and doctors can identify and should be aware of that early signs that maybe more people would recognize that they need to actually get their child or maybe their patient or their student further evaluation rather than just saying well, they will improve, which is what we were kind of told—well, it is early, she is only 6, you know, and we are like, no, no, no. All the other 6-year-olds are reading at a much higher level and her brothers did and we know it. So what should we be aware of, the early warning signs? What can we do?

Dr. SHAYWITZ. That is a good question, and I am so glad you asked it. Children can and must be identified, and the parents are the ones who know the child best. Sometimes schools don't listen to them but parents do know the child best. Dyslexia is a language-based disability. So you can start right away. There may be a delay in talking. As children grow up and go into toddlerhood, and I think Ms. Antie mentioned this, one of the first things—dyslexia results in difficulty getting to the individual sounds of spoken words, so children have to be able to separate out the individual sounds into each individual sound that represents the letter. One of the first times the child has to do that is when they have to rhyme because how do you know cat-mat rhyme. You pull it apart and you look at the "A." So children who have difficulties rhyming, who have difficulty learning the sounds and the letters, that is usually—so that is something that parents can watch for but also pediatricians.

I am a pediatrician, and I must say, pediatricians need to know more. They are really the people who are often the ones that follow the child, know the child, but they don't know enough about dyslexia, and we at the Yale Center are taking actions to educate pediatricians, to involve them more, and we have now actually developed some instruments that can be used for screening at kindergarten and at first grade.

So I have just begun what are the symptoms. I have them in my book, Overcoming Dyslexia, but they should be made part of teacher training. Every teacher shouldn't deny when a parent is concerned but should actually know what dyslexia is, as Max said, knowing what it is about and be able to identify that child and have the support at the school to be able to do that, and that doesn't occur.

I just want to add also that very often—and we have been talking about Charles Schwab and others—dyslexia is an unexpected difficulty, as I showed on the slides. You can have a high IQ and read much below that, and very often, schools, if you are bright, they say oh, well, you know, as Max was told, you are not trying hard enough or you have to work harder. So children and young adults who have high levels of intelligence—and you can test for it and it must be tested for—shouldn't be overlooked as well.

Mr. BUCSHON. Thank you.

86

Dr. SHAYWITZ. Thank you for that question.

Mr. BUCSHON. You are welcome. And I just want to say again, it is not, in my opinion—I have got four kids. Anyone who has children knows, we could tell when she was younger, and we were advocates because we had experience. I would implore everyone to, as a parent, to be an advocate, and the last thing, and then I will yield back, Mr. Chairman, is that anyone who has kids also knows that when you are in kindergarten or first grade, the social status starts to develop of where you fit amongst your peers. It is very, very early, it is much earlier than that even. It is in pre—so to wait can do a lot of damage.

Dr. SHAYWITZ. I just want to——

Mr. BUCSHON. I yield——

Dr. SHAYWITZ. —add that data hot off the presses shows that the gap is there by first grade, and it is just not acceptable to wait and watch. That is waiting for failure. To reinforce what you said, children know who is in the sparrows reading group and who is in the eagles reading group.

Mr. BUCSHON. They do.

Dr. SHAYWITZ. And you can't fool them.

Mr. BUCSHON. You can't.

Dr. SHAYWITZ. Early identification should be mandatory. Teachers have to look for the signs. Children need to be evaluated and they can be and then receive the evidence-based instruction that they require. You can turn an unhappy child into a happier one but if you do it early, you don't have to go through that unhappiness.

Mr. BUCSHON. Thank you, Mr. Chairman. I yield back.

Chairman SMITH. Thank you, Mr. Bucshon. The gentlewoman from Oregon, Ms. Bonamici, is recognized for questions.

Ms. BONAMICI. Thank you very much, Mr. Chairman, and thank you so much to all of our witnesses for being here today. It has been a very informative hearing. I appreciate it.

First I want to start by telling Ms. Antie, I had to go to a meeting during your testimony but I have read your testimony, and you said you were just a mom. That is a very important job, so thank you so much for bringing your story.

I serve not only on the Science, Space, and Technology Committee but I also serve on the Education and Workforce Committee. I don't take off my education hat when I come here. I also don't take off my science hat when I go there. So I am looking at this from both perspectives, and I just want to say, Mr. Brooks, I have many times questioned whether our national overemphasis on standardized tests is inhibiting creativity, and you certainly answered that question for me.

I have had a lot of meetings in my wonderful district in Oregon with some of our decoding dyslexia Oregon parents, and they along with all of you understand that there have been so many advancements in research, in technology. Just look at the Intel Reader, for example. There have been a lot of things that have been done. We find here in the capital that oftentimes the technology is ahead of the policy, and so I wonder if you could—I will start with Dr. Shaywitz. If you could talk a little bit, you mentioned evidence-based programs, so important. So I would like you to focus on what are some of the evidence-based programs and then also I want to have time to talk about the early indicators, and Mr. Brooks, you got great applause to your suggestion that there be the mandatory recognition training for teacher certification but there are a lot of teachers who are already teaching. We need to get to them as well as to the new teachers. So Dr. Shaywitz, could you please talk about what are some of the evidence-based practices that we should be promoting?

Dr. SHAYWITZ. Thank you. I am happy to do that, but I think first, it is important to differentiate evidence-based from research-based. We often hear people saying this program is research-based. Evidence-based means that there is proven efficacy. Research-based simply indicates there are theoretical suggestions but does not provide evidence that the program is actually effective. Evidence-based programs are akin to the level of evidence the FDA requires before medication can be approved for use. Many, many theoretical research-based approaches when tested in the field prove to be ineffective. Our children's reading is too important to be left to theoretical but unproven practices and methods. We must replace anecdotal and common but not evidence-based practices, with those that are proven, that is, they are evidence-based. That goes for programs to teach children to read, for programs for professional development, and programs that colleges of education use when they are teaching future teachers. There are a number of evidence-based programs and they have in common that they reflect the knowledge we know about reading, that they go back to teaching children about spoken words, about pulling the words apart, attaching them to the letters.

In 1998, Congress became very concerned that there seemed to be an epidemic of reading problems and mandated that a National Reading Panel be constituted to investigate what methods and approaches have evidence. I was honored to serve on the panel. The Report of the National Reading Panel, which indicated that the components that are necessary and they include not only phonetic awareness, phonics, fluency but also vocabulary fluency and comprehension.

Ms. BONAMICI. That is great, and I am going to have time to talk about accommodations a bit, but obviously we have heard today and we know that not every child who has dyslexia has involved parents who can dedicate the time and have the ability to advocate for them like we have heard about today, some of these great examples. We think about the lost potential. So not every child can even get to a pediatrician, sadly, so I am a big fan of school-based health clinics, but we have to make sure that we are doing something so that our teachers can play this critical role as early as possible in that diagnosis.

So can we talk about accommodations? I am concerned about both for standardized testing and for, for example, college admissions that may require a second language. I was interested to hear that Representative Brownley's daughter is trilingual, which I think is a huge accomplishment for someone with dyslexia. So are we doing enough in accommodation? Drs. Eden or Dr. Shaywitz or anybody?

Dr. SHAYWITZ. I think that people often misunderstand the role of accommodations. They may say oh, that is a perk, and it is not a perk. In my experience, so many students that I see, I say you need this accommodation respond by saying, "oh, no, I am so afraid if I request accommodations, I will stand out, people will either think I am not so smart or I am being in the system." In truth, dyslexic people often can learn to read fairly accurately but they don't read automatically or fluently, so it becomes very effortful. They have the knowledge but they don't have the time, and so one of the most important accommodations is the provision of extra time, and—both on standardized tests and regular tests but you have to make sure that the accommodations are given in a way that doesn't embarrass the students. So, oh, Johnny, you are going to stay because you need more time, and the individual could just die of embarrassment. There are a number of necessary accommodations, for example, extra time in a quiet room, because when you are not reading fluently or automatically, you are using up all your attention so you can be easily distracted.

We also know that dyslexia causes difficulties with word retrieval. You know what you want to say but your lips and mouth don't form the words so you can be asked "oh, what is that," and it is a volcano but the child will utter "tornado." So teachers have to know that children that have word retrieval difficulties maybe shouldn't be given oral exams, particularly in front of others, and I just—technology can be helpful as well, but right now—and I am very involved in this area—students for high-stakes tests requested from a testing agency are made to go through obstacle after obstacle after obstacle after obstacle, causing many to give up. I know a brilliant woman who wanted to go to law school but she kept getting rejected for accommodation. We are missing out on potential.

Ms. BONAMICI. Lost potential. And my time has expired. I yield back.

Dr. SHAYWITZ. Mine has too.

Chairman SMITH. Thank you, Ms. Bonamici. The gentleman from Arizona, Mr. Schweikert, is recognized.

Mr. SCHWEIKERT. Thank you, Mr. Chairman, and Mr. Brooks, I think you may have accomplished something no one has ever done. That is the first time we have had an outburst of applause, I think, in the Science Committee, and I don't know whether I am creeped out by that or I am overjoyed, but that is a different discussion.

To actually try to get to something a little more serious because we are actually in an interesting inflection point much of this discussion about do we need more science, do we need more research or do we need more carrying out what we know, and Professor Shaywitz, in your opening statement I want to make sure I was listening carefully. You shared that you believe we know substantially the protocols that break through. We just don't carry them out enough. Was I listening appropriately?

Dr. SHAYWITZ. You were, but I would want to add something to that, and again, as what Max referred to, if a child isn't diagnosed and identified, we can know the protocol but they won't receive it.

Mr. SCHWEIKERT. No, I am just—I am one step behind that. I am assuming—let us say I have a student population, and we have our young people, we have them properly diagnosed. The optionality of having a place where they can go where they can receive the type of instructions that——

Dr. SHAYWITZ. Evidence-based instruction, but there is evidence ,too, that the teacher plays an important role. Investigators who have studied different reading programs have found, in the hands of different teachers, the same program can have varying results. So it gets back to the teacher becoming knowledgeable and not only knowing the program but having the toolbox to use to individualize so it is not like a rote back and forth but it is a teacher who knows the child, who knows about reading, knows evidence-based programs and can bring it all together.

Mr. SCHWEIKERT. Well, you are actually right along a policy—it sometimes becomes a policy division and it breaks my heart because it shouldn't be. Louisiana, particularly Arizona, we are in many ways the charter school state in the Nation. In my own community, we have multiple schools that provide specialization for dyslexia, for ADD, for other things, and through the child and the parent, you know, the ability to have that level of parental choice to receive that within our charter-school system, and in some ways it breaks my heart that I know there are many parts of the country where just that even as a discussion is uncomfortable.

So in some ways I was elated to be hearing your words that if we can identify, get the child in a program, great things can happen. Now I need more of the embracing of those programs around the country.

Dr. SHAYWITZ. Yes, absolutely, and I just want to add one other thing. I have had the privilege of visiting the Louisiana Key Academy, and what you see here is again what colleagues have spoken about: it is the whole child that gets the attention and support they needs. We have heard about the pain and the frustration. What happens when you are in a school that totally understands and embraces you instead of pushing you away, you get the reading instruction but at the Louisiana Key Academy, even the phys-ed teacher, I have met him, programs for those students, understands the students, so that they are not told you are not trying hard enough but they feel good about themselves and they learn. So you have to have the whole child in mind.

Mr. SCHWEIKERT. And at one point Massie and I here were geeking out on trying to do the statistical probability of having two doctorates with the same last name, and not related, correct? We almost have the number but we are still working on it. He is the guy who went to MIT so—but within Dr. Eden's discussion, actually a little bit of both of you, we have great technology, we have some great knowledge, but having great data when you haven't built the robust systems to carry it out, and I am hoping that they are embracing your data at his university to carry it out for your information to produce more people like Mr. Brooks who I forgot to bring my book because I wanted an autograph.

And the very last thing, Mr. Chairman. Your book was actually recommended to me by someone who specializes in conflict resolution around the world, and his premise, unlike the movie, was saying look, this guy wrote what he thinks would happen in different cultures, and somehow you broke through with that.

So with that, Mr. Chairman, I yield back.

Chairman SMITH. Thank you, Mr. Schweikert. The gentlewoman from Maryland, Ms. Edwards, is recognized for her questions.

Ms. EDWARDS. Thank you, Mr. Chairman, and thank you very much to our witnesses.

You know, as I was sitting here and I was thinking, you know, to moms and dads, as a mom of a child who has a complex of disorders that are dyslexia, dysgraphia, those things, I remember the challenge of just trying to get both the school system early on to recognize what I knew as a parent when he was a toddler and ready to go into kindergarten so complexities with the school system, then trying to get the health insurance system to recognize that I needed help to be able to pay to make sure that he could— that my son could be tested and fighting with them actually for a couple of years until finally I actually just said you know what? I am not going to pay my mortgage this month because I am going to pay for him to be tested because I couldn't wait another 4 months or go into another school year knowing that I knew that there was a problem and I couldn't get the system really to listen. And the best thing I did was to go to the Lab School here in Washington, D.C., and to have my son tested and then to be able to figure out how to work from first grade on to get him the kind of interventions and tools that he needed to be able to be a success, listening as a parent to educators saying well, here are the list of things that he is never going to be able to do, not be able to read and not be able— he will probably never go to college so maybe you should think about some kind of skill. I had educators tell me that as a parent, and I just said not my child on my watch.

And when I think about parents and the challenges that they have with these various systems to be able to fight for their children, I just want to thank you all for being here today and sharing with us because now as a Member of Congress going into so many classrooms where children are set aside and, you know, they have, you know, some sort of challenge that doesn't allow them to pay attention in the kind of way they need in the classroom but they are struggling and instead they are described as disruptive and unruly or they are not paying attention in the classroom. I think how many millions of children we have around this country and that is what is happening to them, and I want to ask you, particularly as it relates to African American children—and we, you know, focus a lot of attention especially by the fourth grade level where we see those gaps well before then. I know with my son, by the time he was tested going into first grade, the intelligence gap in his—and the performance and reading gap was so significant, it wasn't until I saw the chart that I realized what was happening with him, and so I want to know what it is that we really can do with all of these systems to have them integrate what needs to happen at the earliest level, at the pre-K and kindergarten level so that we don't lose these children through 12th grade. And so any of you who can, you know, share with me, you know, our good doctors about what we can do, because I don't want another parent to have to forego a mortgage payment because the health insurance system doesn't step up so that they can get their child tested.

Mr. BROOKS. Let me say something quickly. I think from an emotional and a social point of view, I think the Chairman at the very beginning of the session already brought up an amazing solution, which is role models, and that is something that the African American community already has the drop on as far as kids feeling good about themselves, and this is something that is very important.

When I was a kid, there were no role models for dyslexia. The only one they knew about was a pole vaulter, which made me think, well, that is great; if I want to spend the rest of my life jumping over things with a stick, that would help, but I think there needs to be an accessible national database of successful people with dyslexia in all the fields. So that is the first thing you can say to a kid. So the first thing when a kid of any background says I am dyslexic, then we say to them, yeah, you and Einstein.

Ms. EDWARDS. Doctor?

Dr. GUINEVERE EDEN. I would just like to go back to this issue of the fact of what we already know and the fact that much of what we know isn't being practiced and implemented and the reality out there, and it goes back to different kinds of information, one of which, which we haven't mentioned yet here but we know from all of you who have spoken about those of you who have children with reading problems in your family but also maybe not just your children but other family members, and that is, if you have dyslexia, the chances that your child will have dyslexia are much, much higher, somewhere around 35 percent, because it is a genetic—multiple of genes are contributing to this. So we know who the children are who are likely to have reading problems when they are very small if they have it in their family. The kinds of tests that tap into children's oral language skills like phonemic awareness that are measured like rhyming the way Dr. Shaywitz just described. Those tests are out there. There are standardized tests that have been out there for years. They are in the classrooms. They could be there. They are not being used. They are not being used because probably somehow the person who should be using them hasn't been taught how to use them and then doesn't know what to do with them and then doesn't know what to do next.

So there is a lot of research, and I think this is the sort of general frustration that we are feeling here is that there is a lot of lower extremities amongst those who are familiar with those tools that they can be used to identify children early and then to teach children early, and teaching children with dyslexia isn't radically different from teaching other children to learn to read. It is many of the same principles. It is just that their delivery has to be different. It has to be in smaller groups. It has to be more systematic, use other techniques to enhance the specific and emphasize specific areas but it is not like these are children who have to learn completely differently. And so what I think what we are hearing here is concern that teachers don't have these tools, and we hear this when you go to a conference like those hosted by the International Dyslexia Association, you hear the dialog between teachers and people who are in research and who do reading research, and the teachers say well, we didn't know that, you know, nobody taught us that at the school and we just don't know that. We didn't know we had these tools to evaluate reading or to help them to become skilled readers, so there is—the biggest problem is really is making these things available and making sure that the people who are using these tools are using them in a way that they were intended to be used so that the children who are they being used for can benefit from it.

Ms. EDWARDS. Thank you, Mr. Chairman. My time is long expired. Thank you.

Chairman SMITH. Thank you, Ms. Edwards. The gentleman from Kentucky, Mr. Massie, is recognized.

Mr. MASSIE. Thank you, Mr. Chairman.

Dr. Shaywitz, I think you helped us all by dispelling a couple myths early on, and you were the first to present one of the myths I think you dispelled that this is centered in the visual cortex. Instead, it is in the center where language is, or as one of your charts aptly demonstrated, this is not an IQ deficiency. What are some of the other myths that you think we should dispel here today?

Dr. SHAYWITZ. I see why he is next to me. I need help.

I think there are a number of myths. One is that dyslexia isn't real, that it doesn't exist. You know, schools will say we don't believe in it, and I always answer that by saying that in religion you can choose what to believe in but dyslexia is scientific and factual so that a big myth is that it is not real.

Another myth is that dyslexia affects only boys and not girls, and that came about, I think, because of what we found, I direct a longitudinal study where we actually tested every child, and we found that, schools did identify many, many more boys. Why? Because boys were annoying the teachers—boys can be boys, and they were chosen by the teacher to be the ones evaluated. Girls who would be sitting very properly but not reading a word were overlooked, so that is another myth.

Another myth is that it is not universal—oh, it is just here and there. Dyslexia occurs in all parts of the world. I wrote a book, Overcoming Dyslexia, and a few weeks ago I was surprised to learn that it had been translated into Chinese. So dyslexia is universal. It affects every culture, every ethnic group, every socioeconomic group and every language system. So that is another myth.

And I think the worst myth is that people who are dyslexic are not smart, and one will hear people like Max and others, it is an unexpected difficulty, so that means we have had such great examples that very bright people can be dyslexic, and I for one don't want to hear anymore, ''well, he is dyslexic but he is smart.'' It goes together. Why should that be a ''but''? I think the belief that people who are dyslexic aren't smart is really one of the most harmful and inaccurate myths.

Mr. MASSIE. Thank you.

Well, let me ask one of the smartest guys on the panel here, Mr. Brooks, my next question then, because he has helped to dispel that myth. I know my kids are a fan of your work. One thing that I wanted to ask you, I know we have focused on children and teaching here and identifying this early but obviously there are adults who never got identified or there are folks like yourself who were identified and the learning was adapted, you know, for your condition, what do you carry into adulthood that you still have to cope with? You know, you mentioned sort of humorously that you weren't going to read your statement, but what are the things you have to cope with still in spite of the care that you received or teaching?

Mr. BROOKS. You know, I would say that most of my books are very research-based, which means that when I do my research, the challenges are, any time there is a new word, any time there is a new phrase so I have to deal with new technologies, new cultures, it is always a challenge, and what I learned in school I still take with me, which is what I do when I have to read is, I listen to the audiobook with the hard copy on my lap, so that way I am taking in the information verbally but then I am underlying everything in the actual book so then when it is over, I don't have to go back to the audiobook, I have the hard copy with me. But research is always tough, and it always takes me longer. So I still struggle with it all the time. In fact, whenever I have a book that is about to come out, I always hire a fact checker to make sure—because my self-esteem is still iffy—to make sure I got everything right. And like in World War Z, I got the weapons systems right, Chinese politics right. There was a sporting-goods store a block away from my apartment, I put it on the wrong street. I got it wrong. So I am always aware.

Mr. MASSIE. This is a question for anybody that cares to answer. To what extent do you think sometimes parents delay the testing because they are worried about a stigma associated with their child for being diagnosed with dyslexia? They want to hope that their kid can stay, you know, at the mainstream or whatever and doesn't need any special attention and the stigma associated with that. I mean, do you think that factors in to the delay in identifying this? I like Mr. Brooks's recommendation that teachers, you know, that is a requirement of their certification but what can we tell parents?

Ms. ANTIE. I will answer this. I never once worried about the stigma my child had. I could care less what they labeled him. As long as they gave me a diagnosis and ways to help him, I don't care what the diagnosis is, and I am sure every parent feels that way.

As far at the stigma, he already had low self-esteem because he couldn't read in front of everybody else. So the fact that he got pulled out for untimed testing didn't hurt his self-esteem any more than it already was. So I don't think the stigma has anything to do with it, personally.

Mr. MASSIE. All right. That is good to hear. Well, my time is expired. Thank you, Mr. Chairman.

Chairman SMITH. Thank you, Mr. Massie.

Let me thank all of our expert panelists today. You have contributed much to the subject, and it is just wonderful to hear all your stories and all your observations as well. I really do think this is one of the best hearings we have ever had, and so appreciate especially your participation, and I would like to thank everyone in the room as well for helping to make this hearing a success.

A final reminder: When you go out into the hall, turn left and join us for lunch and further discussion about dyslexia. Thank you all again for being here, and we stand adjourned.

[Whereupon, at 1:05 p.m., the Committee was adjourned.]

Appendix I

ANSWERS TO POST-HEARING QUESTIONS

ANSWERS TO POST-HEARING QUESTIONS

Responses by Dr. Sally Shaywitz

"The Science of Dyslexia"

Dr. Sally Shaywitz, Audrey G. Ratner Professor in Learning Development, Co-Director, Center for Dyslexia and Creativity, Yale University

Questions submitted by Rep. Julia Brownley, Committee on Science, Space, and Technology

1. In 2013, the state of New Jersey passed a law requiring each school district to screen every student who has exhibited one or more signs of dyslexia, or any other reading disability, no later than his or her completion of the first semester of second grade. Is this a policy you would recommend be implemented on a national scale?

    The key is how the policy will be implemented. As I emphasized in my testimony, it is critical that decisions are evidence–based and that prior to implementing such a policy on a national scale it cannot be left to anecdotal or received wisdom, but it must be based on rigorous scientific evidence. Before any decisions are made – policy makers must demand "Show me the evidence."

2. What, if any, has been your experience in the extent of health insurance coverage of diagnostic testing for dyslexia?

    In my experience health insurance rarely covers diagnostic testing for dyslexia.

3. During the hearing you explained how the gap between typical and dyslexic readers is already present by first grade, as well as the importance of early identification. The Committee heard from a parent witness who explained how her son struggled with rhyming games, even before he learned to read. Does it follow that dyslexia can be identified at an age before a child can read but when they are beginning to recognize or sound out letters? At what point should parents consult an expert or have their children tested if they have concerns about their child's language or reading development?

    Dyslexia reflects a basic difficulty getting to the sounds of spoken language. The earliest clue that a child is at-risk for dyslexia may be observed in the child's speaking, for example, the child's late speaking and the child's failure to appreciate rhymes. In my book, "Overcoming Dyslexia" I list early clues that can be seen in the preschool period. Parents should not wait; they should consult an expert as soon as they notice a child is not progressing. Waiting can be harmful.

4. How much training does the average K-12 teacher receive regarding the academic, psychological, and social impacts often associated with students with learning disabilities like dyslexia? Are educators and other professionals appropriately prepared and trained to address the unique needs of students with learning disabilities, such as dyslexia?

    In my experience this is covered very briefly in teachers' education, usually, as part of a survey course. It is important for teachers to be knowledgeable about reading and dyslexia. At the same time, I urge great caution in selecting programs to train teachers about dyslexia. As I emphasized throughout my testimony, the goal should not be to just add more, "to do something" or to slap on the words "evidence-based" without any real, rigorous evidence that is scientifically valid. My goal is to bring science and education together and that means taking the science, the nature of the evidence seriously. Any program used to teach teachers about dyslexia, to provide professional development to teachers or to teach dyslexic children **must have rigorous scientific evidence that** *dyslexic children who are taught by this program actually improve their reading.* **For example, it is not enough to show that the teacher benefits; there must be evidence that the** *students taught by the teachers using the program actually show significant improvement in reading* **as measured according to rigorous scientific standards .**

Responses by Mr. Max Brooks
HOUSE COMMITTEE ON SCIENCE, SPACE, AND TECHNOLOGY

"The Science of Dyslexia"

Mr. Max Brooks, Author, Screenwriter, Actor

<u>Questions submitted by Rep. Julia Brownley, Committee on Science, Space, and Technology</u>

1. In 2013, the state of New Jersey passed a law requiring each school district to screen every student who has exhibited one or more signs of dyslexia, or any other reading disability, no later than his or her completion of the first semester of second grade. Is this a policy you would recommend be implemented on a national scale?

In response the first question: do I believe that New Jersey's 2013 law regarding mandatory testing for dyslexic children from 2nd grade on should be expanded to a national program. Yes. Completely. I was actually unaware that New Jersey had already taken this step and I strongly believe that their policy should immediately be implemented on a national scale. Diagnosing dyslexic children is a simple, cost-effective means of immediately changing the course of a child's life. As I said during the hearing, the worst part of having dyslexia are the emotional and psychological issues it creates. Simply creating awareness of dyslexia, both in the home at the classroom will go a long way to alleviating these issues and put children on the path to success.

2. What, if any, has been your experience in the extent of health insurance coverage of diagnostic testing for dyslexia?

As far as your second question, what has been my experience, if any, with health insurance coverage of my diagnosis, I would have to say I have had none. Back in the 1980s, as we all know, the world was a very different place for both dyslexia and medical insurance. Given that my mother has passed away, I have no knowledge of how she and my father ended up paying for my testing (although I suspect it was out of pocket). As I said before, I was lucky. My parents had the private resources to pay for my diagnosis. That is a rare case. When I sat before the committee and heard that a member of the United States House of Representatives had to decide whether to pay her mortgage or get her child tested, my stomach turned. Insurance companies need to cover dyslexia testing. That should be mandatory. If that means that my insurance premiums will go up, so be it. It's a small price to pay. If a little extra money up front means the difference between a successful, productive adult, and an unproductive, possibly criminal drain on the system, than I am more than willing to help pay my ounce of prevention.

Responses by Ms. Stacy Antie

HOUSE COMMITTEE ON SCIENCE, SPACE, AND TECHNOLOGY

"The Science of Dyslexia"

Ms. Stacy Antie

Questions submitted by Rep. Julia Brownley, Committee on Science, Space, and Technology

1. In 2013, the state of New Jersey passed a law requiring each school district to screen every student who has exhibited one or more signs of dyslexia, or any other reading disability, no later than his or her completion of the first semester of second grade. Is this a policy you would recommend be implemented on a national scale?

    I would recommend that this policy be implemented on a national scale. Not every parent has the ability to be as vigilant as I have been with my son. Some parents might not even know the signs or symptoms of dyslexia. Teachers; however, should be able to recognize the signs and send the child for testing in the school. This way, every child that needs help with a reading disability will be able to get that help, at an early age, and no child will have to live his/her life without the most valuable gift we can give any child – the ability to read.

2. What, if any, has been your experience in the extent of health insurance coverage of diagnostic testing for dyslexia?

    My experience with my health insurance coverage for diagnostic testing for dyslexia was horrific. I contacted my insurance company after a meeting with my son's teacher and explained that I wanted to have him tested for dyslexia. They had me meet with my primary care doctor, which I did, for a referral to a local therapist for testing.

    I made the appointment with the local therapist the following week. Once we arrived at his office, we were greeted by his assistant. I was handed many forms to complete and I handed her some information that the school provided and some handwriting samples to the assistant and we were ushered into a room in the back. There, a computer screen was set in front of me and I was interviewed, by the therapist, through Skype. The assistant ran a few tests on my son and I received his report with a diagnosis of Dysgraphia in the mail a few weeks later.

    Furious, I called my insurance company after I met with the doctor through Skype. I called again after receiving the diagnosis in the mail. I explained that my son was given a diagnosis without ever meeting the therapist in person. The insurance company explained that they understood my point but they would not be able to pay for another test. They stated that I could appeal their denial for another test. While I wanted to do that just out of principle, each day that passed, my son was falling further and further behind, and without knowing exactly what was wrong, I wasn't able to help him.

I made the decision to have him tested again. I contacted a clinical psychologist and decided to pay $1,800.00, out of pocket, for my son to have a comprehensive psycho-educational evaluation. There, I met with the psychologist, in person, without my son. We returned to the psychologist's office a week later and my son was tested on 12/22/2011. I returned to the office to meet with the psychologist for the feedback meeting on 1/18/2012.

To this day, that insurance company never reimbursed me one cent for the additional test. However, finally knowing how to help my son was worth every penny of the $1,800.00.

Responses by Dr. Peter Eden
HOUSE COMMITTEE ON SCIENCE, SPACE, AND TECHNOLOGY

"The Science of Dyslexia"

Dr. Peter Eden, President, Landmark College

Questions submitted by Rep. Julia Brownley, Committee on Science, Space, and Technology

1. In 2013, the state of New Jersey passed a law requiring each school district to screen every student who has exhibited one or more signs of dyslexia, or any other reading disability, no later than his or her completion of the first semester of second grade. Is this a policy you would recommend be implemented on a national scale?

Screening for dyslexia by first semester of second grade (NJ model)
It is important to recognize the ground-breaking research happening in the field of LD around neuroplasticity and what it means for early signs of dyslexia and/or reading disability. For one, early screening and diagnosis, without recognition of the intervention potential of this new research, has the danger of miring us in over-diagnosis, labeling, and stereotyping without viable solutions. We definitely do not want to recommend a policy on a national scale without considering its broader implications. *Identification – Diagnosis – Intervention* must be considered a package deal.

That being said, if Response to Intervention (RTI) is properly implemented in schools, then early signs of dyslexia should be observable and detected within the established mandate of No Child Left Behind legislation. General screening of all students is part of Tier 1 of RTI, and if implemented as intended, it should be able to identify students who are having difficulty keeping up with the general curriculum, including those with dyslexia. False positives do happen for a variety of reasons, particularly if teachers lack the training to distinguish between symptoms of dyslexia and other reasons why students are not keeping up with grade level tasks.

Early screening and labeling can be a double edged sword. In an ideal world, the earlier the identification and intervention, the better the future outcomes; however, if early identification results in labeling and tracking of students into a platform of low expectations and differential education, then it does not serve its purpose. This concern is very real, particularly in low SES school districts where many parents are not familiar with their rights under IDEA and other state special education laws. We do not want to go back to the times when identification and diagnosis was tied to financial resources and federal funds for schools and resulted in over diagnosis, particular of low SES students.
In addition, implications for early identification can spill over into unforeseen repercussions for those who do not get diagnosed early on. Many high functioning students with dyslexia are not diagnosed until they reach high school or college.

Bottom line, early screening for dyslexia should be should be enacted through existing legislation and policies, and linked to interventions, based on the latest research and technological innovations.

2. What, if any, has been your experience in the extent of health insurance coverage of diagnostic testing for dyslexia?

Experience with extent of health insurance coverage of diagnostic testing for dyslexia

In K-12, diagnostic testing is the school's responsibility, unless families want a second opinion or disagree with the school's evaluation. At the postsecondary level where diagnostic testing is the responsibility of the individual student or their family, this can pose a heavy financial burden. Diagnostic testing for LD is expensive and ranges anywhere from $1500 to $4000 per student. Furthermore, resources for diagnostic evaluation are not always easily available and families may have to travel some distance to get their son or daughter evaluated for dyslexia.

Health insurance coverage resources for diagnostic testing for dyslexia is highly varied across states. Some medical insurance companies will accept and pay for diagnostic evaluation for LD, but others do not. Some medical facilities, like Mass General Hospital (http://www.massgeneral.org/speech/services/treatmentprograms.aspx?id=1435) in MA, have dyslexia assessment centers, but these are few.

The greatest challenge with health insurance coverage where available is that it pays only for the cognitive assessment part of the diagnosis and not the achievement batteries, which show the impact of the condition of academic performance. This means that students still may have to pay out of pocket to get comprehensive evaluation.

3. You testified that as many as 83% of students with learning disabilities in post-secondary education are not obtaining accommodations, even if they were identified and received similar accommodations or support before college. What steps can be taken to facilitate the successful transition from K-12 to college for students with dyslexia?

Steps to facilitate successful transition from K-12 for students with dyslexia (low rate of accommodations for those in college).

Many students when they get to college are reluctant to ask for accommodations because they feel they can make it on their own. And the accommodations and resources are simply not as robust as found in many K-12 educational systems. Successful transition to college should encompass both getting into college and being able to graduate from college. In other words, both admission and retention for students with LD are important.

There are several ways by which successful transition to college can be further facilitated:

1. We need to move away from treating all individual students with dyslexia as a single group. Student from different school districts or varying SES have different pathways to transition. Unfortunately, not all pathways are the same and do not result in the same outcomes.

   Once again, we have a mechanism in place. The IEP and the transition plan of the IEP should be the starting point for successful postsecondary transition. The transition plan should include specific action steps that can help guide students and their families towards postsecondary admission. School counselors sometimes set a low bar for high school students with dyslexia about postsecondary opportunities. This is unfortunate as many students often are not given proper and timely information they need to get into a college that matches their potential and capabilities. Careful oversight of how the transition plan of the IEP is written and executed is key. There are many studies that show that even when students are present at their own IEP meetings, they are intimidated and fearful to speak out. Giving students the opportunity to learn and advocate for themselves can be empowering and can create for better retention in postsecondary settings.

Starting college transition planning early is yet another element of successful transition. Despite regulations, the reality of transition planning for students with dyslexia is vulnerable to variations in policies among school districts, special education teacher training, and administrative oversight. Parents are a key stake holder as well and should have the opportunity to participate, especially for those students with dyslexia who are at risk for not getting into college. Not all students are ready for college immediately after high school. A transition year, especially for those students who need explicit instructions in study skills/learning strategies, is important.
Transition planning and preparation should include strategies for addressing issues related to executive functioning such as self-confidence, self-determination, locus of control, and threat stereotype. These are not the same as study skills, but often result in self-sabotaging behaviors that can details students' postsecondary goals. Another element of successful transition that has gained much significance in recent year is technology competencies and preparedness with ubiquitous technologies in education.

Colleges can do more to facilitate transition by adopting a more universal design for learning (UDL) approach to higher education. Students with dyslexia are often uncomfortable asking for accommodations in college because of perceived stigma and negative stereotyping. At Landmark College, we do not have a separate office for disability services; instead, the concept of accommodations is seamlessly woven into every element of pedagogical practice. All students have the option to receive individualized support through a proactive model of academic advising, content tutoring, and academic coaching. The student support model is a distributed model where academic and non-academic resources are equally engaged in ensuring a quality education for all students. I would venture to suggest that we need more adoption of the Landmark College model by other institutions around the country, adapted of course to institutional goals and mission.

4. How much training does the average K-12 teacher receive regarding the academic, psychological, and social impacts often associated with students with learning disabilities like dyslexia? Are educators and other professionals appropriately prepared and trained to address the unique needs of students with learning disabilities, such as dyslexia?

Training for average K-12 teachers on students with LD.
Currently, the training received by the average K-12 teacher is quite disparate and varies by teacher training program and institution, across the country. Many teacher training colleges have only one course on special education. Such training clearly needs to be expanded. One proposal is to offer online/blended on-the-job training. Training on students with dyslexia is best when it is anchored in authentic circumstances and real life situations. Landmark College faculty, staff and researchers are experts in this area. The College's mission is to change how students learn, teachers' teach and how the public thinks about education. The model of this college should be emulated across the country.

It is time to rethink traditional models of teacher training in light of technologies that can create for virtual engagement with students who learn differently. We need to find better ways to attract the brightest and the best to education, and in particular, education of diverse learners. Currently, there are too many stake holders in teacher training and unfortunately, with not much communication between them.

HOUSE COMMITTEE ON SCIENCE, SPACE, AND TECHNOLOGY

"The Science of Dyslexia"

Dr. Guinevere Eden, Director and Professor, Department of Pediatrics, Director, Center for the Study of Learning, Georgetown University

Questions submitted by Rep. Julia Brownley, Committee on Science, Space, and Technology

1. In 2013, the state of New Jersey passed a law requiring each school district to screen every student who has exhibited one or more signs of dyslexia, or any other reading disability, no later than his or her completion of the first semester of second grade. Is this a policy you would recommend be implemented on a national scale?

Answer: Yes, this and the other dyslexia-related laws that have been put in place in New Jersey are important and encouraging. Early screening and identification means that appropriate intervention can begin sooner. If appropriate intervention is successfully applied, then these children can be prevented from falling behind further and having to face the secondary consequences of reading disability, which are not only academic in nature, but can also manifest as emotional problems.

The New Jersey law could be further improved in two ways. First, screening should occur even sooner, not as late as 2nd grade. Children who are at risk for reading problems can be identified using tests that are quick and inexpensive as early as kindergarten. Second, I believe that when enrolling their child in school, parents should have the opportunity to be asked about family reading history. Parents are typically asked questions about their child's medical history, including allergies. A question about problems with reading amongst their child's first-degree relatives should be included, as it yields information about familial risk. Genetic research studies have shown that the likelihood of a child having reading problems when one of their parents does is around 35 to 40%. Familial risk would provide further valuable information in addition to that gathered through testing of the child.

2. What, if any, has been your experience in the extent of health insurance coverage of diagnostic testing for dyslexia?

Answer: Health insurance companies differ in their policies and coverage. In general, it is my impression that families receive little coverage for assessments, consultations, and treatment of dyslexia.

3. Your research indicates the brains of females with dyslexia differ from the neurological models of the brains of males with dyslexia. How significant are these differences? What implications might they have in determining the appropriate identification, intervention, and teaching methods for men versus women dyslexics?

Answer: Our study, investigating dyslexia in both males and females, is the first to directly compare brain anatomy of females with and without dyslexia (in children and adults). Because dyslexia is two to three times more prevalent in males compared with females, females have been overlooked, and it has been assumed that the results of studies conducted in men are generalizable to both sexes. However, it is known that male and female brains are different in general, anatomically and also in function. It is also known that sex hormones are related to brain anatomy and function, and that female sex hormones can be protective after brain injury. The results of our study demonstrate that researchers need to tackle dyslexia in each sex separately to address the unanswered questions about its origin. This future work will determine to what degree identification and treatment is sex-specific.

4. You testified that as many as 83% of students with learning disabilities in post-secondary education are not obtaining accommodations, even if they were identified and received similar accommodations or support before college. What steps can be taken to facilitate the successful transition from K-12 to college for students with dyslexia?

Answer: I personally did not testify on students with learning disability in post-secondary education. However, as a faculty member of a university involved in teaching undergraduate and graduate students, I believe that a successful transition to and completion of college requires (1) the student's ability to effectively advocate for themselves to ensure they receive the accommodations to which they are entitled, (2) well-trained staff and appropriate provision at the college's academic resource centers, and (3) greater awareness and understanding of learning disabilities amongst the teaching faculty at the college.

5. How much training does the average K-12 teacher receive regarding the academic, psychological, and social impacts often associated with students with learning disabilities like dyslexia? Are educators and other professionals appropriately prepared and trained to address the unique needs of students with learning disabilities, such as dyslexia?

Answer: Teaching reading to students with dyslexia requires considerable knowledge and skill. The amount of training of the average K-12 teacher varies widely in both quantity and quality during early teacher preparation, as well as for later professional development as a teacher. Generally these teachers and other professionals are not appropriately prepared to address the needs of these students. As stated in the International Dyslexia Association's *Knowledge and Practice Standards for Teachers of Reading* (http://www.interdys.org/standards.htm), "Regrettably, current licensing and professional development practices endorsed by many states are insufficient for the preparation and support of teachers and specialists." Even more concerning is that reading specialists and special education teachers often do not know more about proven effective practices than the general education teachers. The biggest concern is that most teachers have not received the training to allow them to recognize early signs of risk of reading problems and to teach students with dyslexia effectively. The courses that lead to certification of teachers and specialists are often found to be lacking in content, especially current research on effective instruction for students at risk for reading problems. To address this problem, the International Dyslexia Association has adopted the *Knowledge and Practice Standards for Teachers of Reading* and is providing accreditation to those university teacher training programs in reading that meet these standards.

Appendix II

ADDITIONAL MATERIAL FOR THE RECORD

WRITTEN STATEMENT SUBMITTED BY MATT MOUNTAIN,
DIRECTOR, SPACE TELESCOPE SCIENCE INSTITUTE
PROFESSOR, PHYSICS AND ASTRONOMY,
THE JOHNS HOPKINS UNIVERSITY

3700 San Martin Drive
Baltimore, MD 21218
VOICE (410)
338-4710
FAX (410) 338-2519
mmountain@stsci.edu

Office of the Director

September 18, 2014

Honorable Lamar Smith, Chairman and
Honorable Eddie Bernice Johnson, Ranking Member
Committee on Science, Space and Technology
United States House of Representatives
2321 Rayburn House Office Building
Washington, DC 20515

Dear Chairman Smith and Ranking Member Johnson:

I would like to offer the information below for the committee's consideration for its hearing on "The Science of Dyslexia."

I have a Ph.D. in Astrophysics and today run NASA's Hubble Space Telescope, but I didn't really learn to read properly until I was about nine years old. Because I'm dyslexic, I subsequently discovered I had learned to read a little differently from most people, and written testimony like this one, still takes work.

As many with dyslexia have experienced, my early school years were confusing. I was classified as "slow" and my parents were advised not to expect too much from me academically, but I was as curious, argumentative and eager to learn as anyone else in my class. It took supportive parents and an enlightened teacher to change my world. I was bribed to read, and when I was about thirteen, I started getting A's and B's for my English essays. Essays were always a challenge, but for the first time I started to get excited about writing. Months later a friend looking over my shoulder at the latest piece I had just gotten back from my English teacher, exploded, *"How did you get a B for that? Your spelling is all wrong!"* This wonderful teacher had decided it was more important I got my ideas down on paper and the arguments well structured than worrying about a few random substitutions of "is" instead of "in," the odd missing word and some spellings that even today would challenge Google.

Looking back I was in the end lucky to have supportive parents and imaginative teachers. They came to recognize there was nothing wrong with my thinking or my memory. I just had to read things a little differently and pay a little more attention to my writing like asking someone else to check it (something I still do today). Of course the other breakthrough for me was finding I enjoyed math and science, which involved solving puzzles and didn't require long essays. However, I suspect my handwritten lab books were somewhat challenging for the uninitiated teacher. In my professional life word processors arrived at the right time. Though as the early ones used to "beep" when you typed a misspelt word, it wasn't unusual for an office mate, or later my wife, to lean over and switch off my computer's speaker when I was in the midst of an intensive piece of writing. Many people today deplore the rise of programs like PowerPoint and Keynote. For me these were the breakthrough communication tools. I could now build my arguments with pictures, diagrams and few words.

I am often asked today whether I also think differently because of how I learned to overcome my dyslexia. Of course I don't really know how other people think. I do know that my parent's bribery worked, and I learnt to read by consuming science fiction books with complex stories: I read the entire Foundation Series by Isaac Asimov in a month. It's the plot and storyline that I soaked up and remembered, not how it was written. Today if I'm reading a document quickly, I absorb the arguments, jump backwards and forwards between paragraphs, but don't always notice good writing, until it's pointed out to me. I have to decide "to read" if I'm going to appreciate a good book or article, or someone asks me to read something they have written. My colleagues today have gotten used to my emails where occasionally they will find an "is" substituted for an "in," "quiet" swapped for "quite" or a preposition just missing. I still have to think about how to spell "bureaucracy," at least to enough precision where my laptop or smartphone can take over. I'm still embarrassingly bad at remembering names though never seem to forget a face. I am good at absorbing and synthesizing diverse ideas and data, just not very good at writing it down. I do like to use a whiteboard or a pad of paper to sketch out ideas and concepts and always think a presentation that doesn't make good use of pictures or well-crafted diagrams is a wasted opportunity.

To run the Hubble Space Telescope and its Science Institute of six hundred scientists, engineers and administrators, you have to be able to listen to smart people (some with Nobel Prizes), synthesize ideas quickly, make decisions and communicate effectively; always remembering Bertram Russell's maxim, *"the problem with communication is the belief it has occurred."* None of these skills have anything to do with dyslexia, though I am perhaps rather more tolerant of people's differing communication styles and approaches. I will admit to being a little intolerant of long ponderous documents. The lesson I draw from my "disorder" is that it took a different path to reach where I am today, but that journey was no less interesting and required picking up some useful skills along the way. However, like many of other journeys we take through life, to get the privilege of running something like the Hubble Space Telescope, it takes a very supportive family, good teachers and mentors, and most importantly, a personal belief that most things are possible with hard work.

Sincerely,

_____

Matt Mountain
Director, Space Telescope Science Institute
Professor, Physics and Astronomy, The Johns Hopkins University